T0331052

Methods in Human-Animal Studies

This timely book provides a methodological guide for how to conduct and theorise research in human-animal studies. In response to critiques of the anthropomorphic slant to human-animal research and the increasing political relevance of animals in contemporary environmental debates, this book emphasises methods which bring to light the animal side of multispecies encounters.

Drawing from the interdisciplinary strength of human-animal studies, this book contains contributions from practitioners and scholars working in sociology, anthropology, ethology and geography. Each chapter uses a case-study approach to present a theoretical framework and empirical application of cutting-edge methods in human-animal studies, from creative writing in multispecies ethnographies to visual methods like videography and body mapping. Organized in three parts – theorizing; collaborating; visualizing – the book equips readers with methodological tools to conduct human-animal studies research attentive to animal lives. Furthermore, chapters reflect on the opportunities, limitations and ethical considerations of research that seeks to understand our more-than-human worlds.

The book is aimed towards undergraduate and graduate students in human-animal studies and scholars investigating human-animal relations. It will also be of interest to practitioners and policy-makers who engage with conservation, wildlife management or the human-animal interface of urban and regional planning.

Annalisa Colombino holds a PhD in human geography from the Open University. She now works at Ca' Foscari, University of Venice, Italy. Her research applies cultural geography to topics such as place-marketing, commodification and branding processes, the geographies of consumption, the bioeconomy and the diverse economy. Most recently, she has been working at the intersection of economic geography, animal geographies and the geographies of food adopting a more-than-human approach inspired by biopolitical thought. Her interdisciplinary research is published in Italian, English and German.

Heide K. Bruckner is a researcher and lecturer currently working in the Department of Geography at the University of Graz, Austria. In her scholarship, she focuses on the promises and challenges of "alternative" food systems in the Global North and South. In particular, she is drawn to embodied, visceral geographies that capture the intersection of personal experience and more-than-human political ecologies. Her diverse research projects span the globe, from the United States and Latin America, to western Europe and the South Pacific.

Routledge Human-Animal Studies Series
Series edited by Henry Buller
Professor of Geography, University of Exeter, UK

The new *Routledge Human-Animal Studies Series* offers a much-needed forum for original, innovative and cutting-edge research and analysis to explore human–animal relations across the social sciences and humanities. Titles within the series are empirically and/or theoretically informed and explore a range of dynamic, captivating and highly relevant topics, drawing across the humanities and social sciences in an avowedly interdisciplinary perspective. This series will encourage new theoretical perspectives and high-light ground-breaking research that reflects the dynamism and vibrancy of current animal studies. The series is aimed at upper-level undergraduates, researchers and research students as well as academics and policy-makers across a wide range of social science and humanities disciplines.

Animals, Anthropomorphism, and Mediated Encounters
Claire Parkinson

Horse Breeds and Human Society
Purity, Identity and the Making of the Modern Horse
Edited by Kristen Guest and Monica Mattfeld

Immanence and the Animal
A Conceptual Inquiry
Krzysztof Skonieczny

The Imaginary of Animals
Annabelle Dufourcq

Winged Worlds
Common Spaces of Avian-Human Lives
Edited by Olga Petri and Michael Guida

Methods in Human-Animal Studies
Engaging With Animals Through the Social Sciences
Edited by Annalisa Colombino and Heide K. Bruckner

For more information about this series, please visit: www.routledge.com/Routledge-Human-Animal-Studies-Series/book-series/RASS

Methods in Human-Animal Studies

Engaging With Animals Through the Social Sciences

Edited by Annalisa Colombino and Heide K. Bruckner

Routledge
Taylor & Francis Group

LONDON AND NEW YORK

First published 2023
by Routledge
4 Park Square, Milton Park, Abingdon, Oxon OX14 4RN

and by Routledge
605 Third Avenue, New York, NY 10158

Routledge is an imprint of the Taylor & Francis Group, an informa business

British Library Cataloguing-in-Publication Data
A catalogue record for this book is available from the British Library

ISBN: 978-1-138-49751-1 (hbk)
ISBN: 978-1-032-50508-4 (pbk)
ISBN: 978-1-351-01862-3 (ebk)

DOI: 10.4324/9781351018623

Typeset in Times New Roman
by SPi Technologies India Pvt Ltd (Straive)

Contents

Figures

Concept boxes

Contributors

Annalisa Colombino holds a PhD in human geography from the Open University. She now works at Ca' Foscari, University of Venice, Italy. Her research applies cultural geography to topics such as place-marketing, commodification and branding processes, the geographies of consumption, the bioeconomy and the diverse economy. Most recently, she has been working at the intersection of economic geography, animal geographies and the geographies of food adopting a more-than-human approach inspired by biopolitical thought. Her interdisciplinary research is published in Italian, English and German.

Heide K. Bruckner is a researcher and lecturer currently working in the Department of Geography at the University of Graz, Austria. In her scholarship, she focuses on the promises and challenges of "alternative" food systems in the Global North and South. In particular, she is drawn to embodied, visceral geographies that capture the intersection of personal experience and more-than-human political ecologies. Her diverse research projects span the globe, from the United States and Latin America, to western Europe and the South Pacific.

Katrina M. Brown is a senior researcher and documentary filmmaker at the James Hutton Institute in Craigiebuckler, Aberdeenis (UK). She has decades of experience in researching animal-human relations – and especially embodied practices of governance – in a variety of areas including conservation, recreation, agriculture and animal disease. Katrina and colleagues have been at the forefront of innovating mobile video ethnography – notably employing go-along wearable cameras - to understand more-than-human relations as practices situated geographically, emotionally and in sensation and movement.

Séverine van Bommel is a senior lecturer with the School of Agriculture and Food Sciences of the University of Queensland in Australia. Her research and teaching focus on the negotiation of meaning in natural resource management (NRM) and its performative effects. Recently, she has become passionate about a more-than-human participation in research, including multispecies ethnographies as ways to reconfigure the politics of life in the worlds that we inhabit.

Susan Boonman-Berson is an animal-geography researcher and speaker at "Bear at Work", Nijmegen, the Netherlands. Her primary interest is on cohabitation between humans and wild animals with expertise in wildlife management practices and human-wildlife conflicts. She is especially passionate about how to think along with wild animals and treat them (more) symmetrically as participants in research, as well as in wildlife management practices.

Darcie DeAngelo is an assistant professor of anthropology at the University of Oklahoma (USA). Her area of focus is on landmine detection industries in Cambodia, especially those that work with animal detection aids. She is dedicated to engaged studies and has conducted research in diverse fields from public mental health disparities to international policy. She also produces public humanities exhibitions where she troubles the boundary between art and anthropology for a wide range of audiences.

Emily Elsner-Adams holds a PhD in geography from Lancaster University. She is an independent researcher, lecturer and consultant in Zürich, Switzerland. Her research interests include human-nature relationships, sustainability, non-profits and social enterprises, and impact measurement and evaluation.

Hannah Jaicks-Ollenburger is an interdisciplinary scientist who specialises in using psychology to develop stronger avenues of communication and inclusion amongst the public, decision makers and scientists around issues of conservation and sustainable development. Hannah received her Ph.D. in Environmental Psychology from the Graduate Center of the City University of New York (USA). She currently works as a Special Assistant to the CEO at Convergence Center for Policy Research.

Erin Jones is an animal studies scholar and dog behaviour professional who focuses on shifting expectations of the human-dog relationship to a more dog-centred perspective, challenging human-centric concepts of autonomy and consent. Erin is currently finishing her PhD at the University of Canterbury, New Zealand Centre for Human Animal Studies and runs Merit Dog Project, an educational platform for dog caregivers.

Owain Jones is a cultural geographer and became the first Professor of Environmental Humanities in the UK in 2014 at Bath Spa University and was deputy director of the Research Centre for the Environmental Humanities from 2016 to 2019 (now emeritus). He has published over 80 scholarly articles on various aspects of nature–society relations and four books about animal becoming, place and landscape.

Nicolas Lainé is a social anthropologist and research fellow at the French National Research Institute for Sustainable Development (IRD). A specialist in environment–society relations, he has conducted several projects in India, Laos and Thailand. He has published numerous articles on human-elephant relations, including a book on the subject: *Living and*

Working with Giants. A Multispecies Ethnography of the Khamti and Elephants in Northeast India (2020).

Siobhan Maderson has a PhD in Human Geography from Aberystwyth University. She is a senior lecturer at the Graduate School of the Environment at the Centre for Alternative Technology (UK). Her research interests include Traditional Environmental Knowledge, pollinator well-being, food security and the role of multiple evidence bases in environmental governance.

Jocelyne Porcher is the research director at INRAE (France). Her research focuses on the working relationship between humans and animals. She is the author of several books on the subject, including *The Ethics of Animal Labor. A Collaborative Utopia* (2017) and *Animal Labor. A New Perspective on Human-Animal Relations* (2019).

Nik Taylor is a critical and public sociologist whose research focuses on power and marginalisation expressed in/through human relations with other species, informed by critical/ intersectional feminism. Nik is currently the co-director of the New Zealand Centre for Human-Animals Studies at the University of Canterbury in Aotearoa (New Zealand), where she also teaches courses for the Human Services and Social Work program at the University of Canterbury, New Zealand.

Acknowledgements

This book project would not have come to fruition without the support we received along the way. Firstly, we would like to acknowledge Jan-Erik Steinkrüger and Lisa Jenny Krieg for their significant contributions at different stages of the project, shaping both the book's initial conception and its final directions. We would like to thank the book's many authors for their insightful chapters as well as for their patience with the project. Mara Miele and Owain Jones provided critical feedback for the introduction through their close reading and helpful feedback. We are also very grateful to Prachi Priyanka, our supporting editor at Routledge, for her guidance along the way. Finally, we acknowledge that the knowledge and findings generated by/through these chapters is only possible thanks to the nonhuman participants who share their worlds with us. For them, and our interconnections, we are grateful.

1 Hidden in plain sight

How (and why) to attend to the animal in human-animal relations

Annalisa Colombino and Heide K. Bruckner

> Animals surround me right now as I write these words: Inside are three cats; sculptures of elephants [...]; photos of cheetahs [...] and a painting of coyotes. Pieces of animals decorate nearly every room (all found!) – bird nests, a porcupine quill, bison fur, [...] the skeletal mouth of a sea urchin. Outside there are butterflies, a huge spider that lives by the porch light, mosquitoes, [...] and the neighbourhood bully cats. Furthermore, there is milk and cheese in the refrigerator, cat food made of cows, chickens, turkeys, salmon, and tuna, honey, leather shoes, a leather softball glove, and household products that have been tested on animals
>
> (Urbanik 2012, p. xi).

> Nobody cares about animals
> (Academic expressing scorn at human-animal scholarship).

> I do not see animals in the fridge
> (Academic laughing with derision).[1]

The material and symbolic presence of animals at once surrounds and alludes us. While many argue that the social and environmental realities of our contemporary world call us to reconfigure how we interrelate with animals in less exploitative ways, doing so first requires our careful attention to how animals emerge, and what their needs may be. Questions about epistemology, methods and methodology quickly arise – how do we begin to understand human-animal relations, the presence of animals, the subjectivity of nonhumans and our interdependence on this planet? This book tackles these challenging questions head-on, as we present a series of interventions from the social sciences which discuss methods that investigate specifically the *animal* side of human-animal relations.

The recognition of human-animal interdependency is a matter of fact and concern at the core of current research endeavours, in and beyond the social sciences. The study of human-animal relations (hereafter HAR) and cross-species interactions, both in proximity and at a distance, is becoming crucial for informing policy and legislation intended to protect the environment, socio-economic wellbeing and public health, whose more-than-human character is increasingly recognised by global research programmes such as One Health (which explores the interface of human-animal-ecosystem interaction) and the Planetary Health Alliance (which looks into how humans disrupt life on earth).[2]

DOI: 10.4324/9781351018623-1

The importance of doing research at the human-animal interface is further exemplified by the current Covid 19 pandemic which, accordingly to the most supported hypothesis, is likely to have been triggered by the proximity of consumers and wild animals sold in wet markets (Xiao et al., 2021; see also Quammen, 2012). The human-animal interface, underpinned by exploitative human-animal relations, is indeed "where" zoonotic diseases originate (Magouras et al., 2020). Scientists also demonstrate that the human interference with the environment, including through global warming, encourages the mobility of animals searching for more suitable habitats. As human influence continues to grow, encroaching upon and shaping animal habitats, species who had previously never interacted are now confronting one another. Thus, many researchers predict that future viruses' transmission across species is likely to increase in intensity and frequency (Carlson et al., 2022), causing further socio-economic and health problematics across the globe. The Covid 19 pandemic and other zoonotic diseases epitomise the fact that the ways in which humans relate to animals does matter in shaping the more-than-human world we share.[3]

Yet, as the Urbanik quote above (2012) notably emphasised, in our everyday routines the lives and deaths of animals are ubiquitous but go unnoticed. Animals are hidden in plain sight. The invisibility of animals and HAR are problematic from a scholarly point of view, for our everyday lives and current policymaking. They are also a common trope of human-animal studies (see, notably, Berger, 1980) and, as the derisive, opening quotes above suggest, also a matter of fact in human societies: the framing of animals as irrelevant, inferior, unknowable, abyssally different from humans – reflecting the long tradition in Western thought and practice of separating humanity from nature – still looms at large. Despite the explosion of studies in HAR (which we discuss in the following sections) that challenge human exceptionalism and dismissive views of animals, instrumental views of nonhuman creatures are embedded in powerful areas of STEM and productivist-oriented disciplines (e.g., agricultural and livestock sciences, mainstream economics and wildlife management). These fields are linked to our dominant economic systems and thus underpin sectors which support everyday human life, including food, drug and cosmetics production and consumption, but also tourism and the pet economy, for example. So much of our contemporary life is built upon animal exploitation (see Shukin, 2009; Nibert, 2013). Yet, for many people, within and beyond science, to start thinking differently would call into question how key areas of everyday (economic) life currently operate and, relatedly, the very ways in which animals are seen – or not seen.

The poor visibility of HAR and the contingencies of the present urge us to intervene in research, practice and policymaking that render animals' presences and needs – and our co-existence – visible and palpable. The imperative does not simply concern the inclusion of animals in our existing research and policy endeavours; it is instead about reconnecting with worlds of nonhuman creatures to recognise humans and animals' vital and deadly interdependence. To learn to coexist in more careful ways, it is imperative we build a profound social and political awareness of the interconnection that ties human and nonhuman forms of life together on this damaged planet. In other words, if we do not see nor understand the connection, why should we care?

For many, (not) relating to animals' lives and deaths has become a dominant, unthought, embodied habit that contributes to further concealing the relevance of nonhuman forms of life. When we encounter animals in diverse contexts, we often perceive them fundamentally through historically and socially constructed hierarchies, as creatures variously categorised and characterised as food, pets, pests, predators, commodities, numbers, monstrous, cute, etc. But how can we bring to light their points of view and experiences so that we – as humans with an obligation of stewardship towards other species – can think about solutions that improve our connected lives? How can methods, primarily drawn from the social sciences, be adapted to more-than-human research practices? How can these methods be combined with knowledges from other fields such as ethology and biology, or with the expertise of practitioners working with animals, to offer sound understandings of human-animal relations grounded in and emerging from specific places, even as the power dynamics favour humans? What can we learn from researchers, scholars, practitioners in the practical and theoretical sense of how to conduct such research?

This book is about offering some answers to these questions. It is about finding ways to magnify and explore the hyphen in "human-animal relations" as the key site to glimpse, recognise, reconnect and understand humans and animals' interdependence. This is a crucial step if we are to consider the points of view and needs of those creatures who have for too long been hidden, marginalised and silenced, and include them more carefully into our everyday lives and policymaking. This book is also a critical intervention for guidance regarding the methodological aspects of human-animal studies (hereafter HAS), discussing opportunities and limitations of qualitative methods that centre the experiences, interests and needs of animals from a more-than-human perspective. In offering these contributions on how to access the animal side of human-animal relations, both practically and theoretically, this volume responds to calls to move beyond human-centred research strategies and engage more intentionally with nonhuman animals' presences, agencies, subjectivities and social lives. The book thus takes seriously the material, symbolic and affective presence of animals by paying close attention to how we methodologically conduct research.

In this introduction, we highlight how our book builds on existing debates in the field by first providing a brief overview of HAS and its methodological challenges. We then argue the fragmentation of methods in HAS research weakens its scholarly and political aims. Afterwards, we outline the major contributions of each of chapter. The introduction ends by reviewing important limitations of the book, as we suggest future scenarios and ways forward for methods in HAS.

Human-animal studies: a brief overview

> In so far [as the behaviour of animals is subjectively understandable] it would
> be theoretically possible to formulate a sociology of the relations of men [sic]
> to animals, both domestic and wild. Thus, many animals "understand" com-
> mands, anger, love, hostility, and react to them in ways which are evidently
> often by no means purely instinctive and mechanical and in some sense both
> consciously meaningful and affected by experience.
>
> (Weber, 1947, p. 104)

Max Weber, a central figure in the development of sociology, acknowledged
that animals could play a role in sociological analysis. Despite this openness
to a sociology of human-animal relationships, animals were largely ignored
by early 20th-century sociology and cognate disciplines (Sanders, 2006).
Animals' lack of speech seems to have discouraged scholars in the social
sciences to take them seriously (Buller, 2015, p. 375).

In the last 40 years or so, HAS have proven to be an expanding field of
knowledge production, formalised through the establishment of peer-reviewed
journals, working and speciality groups, academic centres and curricula.[4] The
name of this field remains a matter of debate. While scholars in the natural
sciences seem to prefer the phrase "human-animal interaction studies" and
others in the social sciences and humanities "animal studies" (e.g. Kalof,
2017; Calarco, 2020), we adopt HAS as it better represents the label commonly used in the social sciences. More precisely, we see HAS as a contested
multi- and transdisciplinary field; evolving as it is animated by diverse
researchers who prefer to emphasise their shared interest in exploring the
hyphen – i.e., the wide spectrum of relations interconnecting animals and
humans – by using different labels that communicate distinct theoretical and
political approaches. While political activism and animal advocacy are at the
core of critical animal studies (CAS hereafter; see Pedersen, 2011 and Taylor
& Twine, 2015 for an overview of CAS), scholars in anthrozoology do not
take an explicitly political stance because it would contradict claims of neutrality or objectivity incorporated in their research.[5] We assert, however, that
scholarship delving into the animal side of HAR is always political, explicitly
or implicitly, for two main reasons. First, exploring HAR from the perspectives of animals involves an ethical imperative to improve the lives of other
creatures. Second, as we argue later, methods, in and beyond the social
sciences, are performative: they help to intervene in reality and create new
possibilities (Law & Urry, 2004).

Generally, contemporary HAS might be divided into two main focus areas,
depending on which side of the relations scholars primarily consider: the
human side of HAR and/or the animal side of HAR. The first has been the
domain of early studies in the social sciences to understand how humans
have placed animals in specific categories, which shaped how animals were
treated (as property, pets, pests, etc.; see e.g., Wolch & Emel, 1998). Before the
emergence of HAS, the focus on the animal side of HAR had traditionally
been investigated by disciplines such as ethology. More recently, the social

sciences have engaged in HAR through focusing on how place, power, history and technologies shape specific HAR, and by interrogating the perspectives of animals themselves.

As discussed by Shapiro's (2020) most recent overview of the field, historically HAS emerged in the 1970s Minority World[6] in response to increased public outcry over endangered wildlife and the mistreatment of animals in factory farming and laboratories.[7] Work in philosophy (Singer, 1975), agriculture (Harrison, 1964), animal experimentation (Ryders, 1975) and ecofeminism (Gaard, 1993) provided the basis for HAS as an academic field of its own. Shapiro (2020) reconstructs the trajectory of the field into four waves, which we briefly review here. The first wave, in the 1980s, marked the beginning of the "animal turn" with the establishment of the Tufts Center for Animals and Public Policy (North Grafton, Massachusetts, USA) in 1983, and the launch of the first journal in HAR, *Anthrozoös: A Multidisciplinary Journal of the Interactions of People and Animals* in 1987. Early scholarship focused on the human side of HAR, adopted quantitative methodologies and identified how a multiplicity of HAR are beneficial to humans, while animals were largely exploited and mistreated. The second wave emerged in the 1990s when more disciplines joined the field adopting qualitative methods to investigate the human side of HAR, with an interest in the power relations, discourse and practices affecting the lives and deaths of animals. The third wave, in the 2000s, was influenced by postcolonial studies and ecofeminist thought (Fraiman, 2012; see also Deckha, 2012), as well as cultural studies and philosophical approaches inspired by Derrida, Foucault and Heidegger (see Calarco, 2008; Wadiwel, 2015; Wolfe, 2003). Yet, as Shapiro notes, this third wave, from Derrida (2002) onwards, has been primarily concerned with deconstructing and theorising animals, rather than engaging with "real animals" (Haraway, 2008, p. 21) and their experiences. Works fuelling the fourth wave, in the 2010s, criticise how scholars in the social sciences and humanities have investigated animals primarily through social constructivist approaches. Recent scholarship in HAS draws from actor-network theory (Latour, 2005) and the rise of new materialism. From these conceptual orientations, key notions (e.g., affect, becoming, performativity; see also the concept boxes in this book) open up the possibility to explore the animal side of HAR, as they highlight the fleshy and visceral (rather than rational) registers of both humans and animals. Importantly, these intellectual frameworks extend agency to animals by reconfiguring it in more-than-human terms, as unrelated to intentionality, and as a matter of relations rather than inherent capabilities and rationality. They also diminish the importance of speech in keeping together "society" (Shapiro, 2020, p. 817), and allow the mobilisation of other-than-talk methods to investigate the animal side of HAR. This book partakes in this fourth wave, as our collected chapters discuss methods which provide insights into the animal-side of HAR.

In sum, building on the last four decades or so, more-than-human theoretical insights have facilitated our understanding of the multiple ways in which nonhumans act with, and alongside humans, compose and animate the world. Such perspectives have been successful in destabilising and challenging humanistic and anthropocentric modes of knowledge production. They have dignified nonhumans, animals included, as quite palatable "objects" of academic inquiry. Almost, but not quite.

Working with "the anathema": exploring the animal side of human-animal relations

> Within a social science discipline like education, where it is axiomatic that our core business is to investigate human learning or the discursive practices and/or materials that guide and enable this learning, more- than-human research practice seems like an anathema
>
> (Pacini-Ketchabaw, Taylor & Blaise 2016, p. 149).

> Academic A: As I am interested in animals... Academic B interrupting A: yeah, but don't say it to your new colleagues or in class!
>
> (Dialogue overheard at meeting).

> It is impossible to know animals' perspectives!
>
> (Exclamation of both shock and certainty, overheard at academic meeting).[8]

Even today, we (as HAS scholars, while presenting at conferences, talking with colleagues, and writing funding proposals) confront the above reactions from academics within and beyond the social sciences who find the investigation of the animal side of HAR amusing, enraging, irrelevant or impossible. Such responses do more than frustrate; they reproduce humanistic and

---

Concept box 1.1: Posthumanism, anthroparchy and HAS

Posthumanism is an intellectual framework which challenges the notion that humans possess sole agency in shaping our world. Emerging in the 1990s, it draws on academic threads spanning from poststructuralism, postcolonialism and feminism, to new materialism as inspired by scholars such as Haraway and Latour. Posthumanist thought orients us towards four main interrelated directions: un-doing binary thinking, overcoming anthropocentrism, considering the agential properties of (nonhuman) matter and re-thinking the role of humans in making the world (see Colombino & Giaccaria, 2021).

  Perhaps the most important intellectual trajectory of posthumanism concerns decentring humans to conceptualise a world beyond anthropocentrism. Such a move also implies challenging anthroparchial structures. Anthroparchy, literally meaning "human domination", is an expression coined by Calvo (2008) to highlight how we do not simply live in societies that discriminate against humans based on social factors (e.g., gender, race, dis-ability), but also against animals by mobilising the category of species. Anthroparchy is a set of discursive and material practices which establish a species hierarchy with humans on top. Within this framework, the higher positioning of humans is used to justify the exploitation of other species relegated to lower positions.

Attempts at overcoming anthropocentrism and anthroparchy require setting aside "human supremacy" in our analyses of the world to let other humans' and other-than-humans' voices, perspectives, presences and agencies emerge. This means "making space" for other-than-human entities, beings and forces, understood as active parts of the planet's histories and geographies.

The challenges for posthuman thought and research practice are numerous and include: developing strategies for re-thinking and remaking the world, finding approaches that avoid putting things and people into rigid containers, thinking through the complicated interconnections, rather than separations, of animals (humans included), natures and technologies, analysing the active presences and agencies of nonhuman forces, beings and elements, and combating practices of discrimination and exploitation based on binary modes of approaching the world.

One strategy to overcome binary thinking and, at least in part, anthropocentrism and anthroparchy, might be applying theoretical frameworks which do not essentialise, fix or capture the world and its dwellers, human and otherwise, and which, importantly, open up novel ways of becoming. Examples of these ideas include: Haraway's metaphors of the cyborg and naturecultures, which point to the hybrid nature of bodies and the world as socio-technical and socio-natural compositions; and Latour's ideas of network, actant and symmetry, which emphasise agency that is spatially distributed (not inherent in bodies, human and otherwise) and posit that diverse entities become agents when enrolled in specific relations. Importantly, such notions are found in philosophies of becoming rather than being, of movements and emergences rather than essences, as elaborated by radically non-anthropocentric philosophers such as Gilles Deleuze and Felix Guattari (for a glossary of posthuman terminology see Braidotti & Hlavajova, 2018).

Much work in the wake of posthumanism interrogates the place of technology in our understandings of the world in which we live. HAS is also often described as being influenced by posthuman thought. While it can be argued that approaches like actor-network theory and science and technology studies have influenced the development of HAS, it could also be suggested that the emergence and establishment of HAS in the social sciences and humanities has contributed to strengthening posthumanism. We argue the latter in the introduction of this book (see also Colombino & Giaccaria, 2021), as HAS demonstrates a long-standing commitment to decentring the roles and influence of humans to instead consider the affective agencies and fleshy materialities of other animal creatures in world-making.

Finally, we want to note that different scholars use the term "posthuman" but with varying meanings. For a clear and accessible discussion of the differences and intersections of strands of thought often conflated under the umbrella term "posthumanism", such as new materialism, transhumanism, anti-humanism and object-oriented ontology, see Francesca Ferrando's 2019 work.

anthroparchial structures (Calvo, 2008; see Concept box 1.1 on posthumanism, anthroparchy and HAS). However, they stem from at least five deeply rooted ideas and beliefs, which beg for some clarification. First, they originate from the illusion that we can precisely know what humans think, feel and experience, perhaps because humans share speech and morphological similarities. Yet, ever since the crisis of representation and of the pursuit of "objective" knowledge, social scientists generally agree that methods create "artificial" contexts, which enable researchers to only partially explore what humans might think and feel. Interviewing, the primary talk-based method in qualitative research, cannot mirror what really people think or feel, but instead provides a means through which researchers can investigate how humans use language to represent themselves, their memories, and experiences – which scholars in discourse analysis, for example, have long argued.[9] Contemporary social science research in HAS draws on abundant work which posits that scholars can never achieve an objective knowledge which mirrors reality. Many researchers in HAS, in fact, understand their work as positioned, more-than-representational and performative. That is, knowledge produced through research practice always emanates from our situated and partial viewpoints, is framed within specific social and theoretical orientations and thus arises within power-laden, embodied research contexts. It is also the outcome of the mobilisation of methods specifically designed to produce data able to shed light on the particular research questions which we have formulated on the basis of (our analysis of) gaps in research. The knowledge presented at the end of a research process is, therefore, a highly mediated process, shaped by multiple factors and is only partially representative of specific realities (see Haraway, 1988) – including the realities emerging as we approach the animal side of HAR. Relatedly, for many human-animal scholars, as this volume illustrates, such realities do not exist a priori, but come into being as we conduct our investigations. Methods are thus understood as performative (Law & Urry, 2004; Miele & Bear, 2022, 2023). They serve a double purpose: first, they create reflexive knowledges, which are perhaps modest but more transparent, intellectually honest and, thus, sound and credible. Second, and importantly for HAS, by intervening in reality, methods bring to light novel possibilities for HAR in practice, politics and research. In sum, the methodologies discussed in this book are not simply about using specific methods to "better observe what is out there". These are also about creating (different) more-than-human worlds which reconnect with animals and carefully consider their perspectives. Understood as performative, methods are thus key to the ethical, pragmatic and political pursuit of achieving more sustainable and respectful human-animal relations. Yet, while the performativity of methods and the partiality of situated knowledge are widely accepted in international debates on qualitative methods in human-centred research, such notions do not yet seem to apply to the worlds of animals.

Second, the impossibility and irrelevance of studying the animal side of HAR is based on consolidated beliefs that nonhuman creatures are profoundly unknowable because of their, supposedly, abyssal difference from humans. Despite abundant work in cognitive ethology and animal welfare sciences that has proven otherwise, animals have long been presented as lacking language, minds, logos, emotions, subjectivities or ethics. Third, and relatedly, the belief that studying the animal side of HAR is irrelevant originates in the Western myth of human exceptionalism embodying a very powerful ideology of human supremacy which, in turn supports animals' normalised expendability as they are constructed as mere numbers, commodities or property. Fourth, scepticism towards studying the perspective of animals is reinforced by the idea that researchers untrained in disciplines such as ethology or animal biology, for example, are destined to fall into the trap, and unforgivable error, of anthropomorphism.

Anthropomorphism involves attributing to animals the characteristics, behaviours, states of minds and emotions which are distinctive to humans, and which scientists have not proven to pertain to specific, other animals (for example, the possibility that fish and insects feel pain is still matter of debate; cf. Gibbons et al., 2022; Rose et al., 2014). However, as Bruni et al. (2018) aptly emphasise, since Morgan's canon (1894), the opposite of anthropomorphism – failing to recognise that some animals have cognitive and emotional traits common to humans – is a mistake without a name only recently identified as "anthropodenial" by de Waal (1999) and "anthroporenalism" by Cartmill (2000, in Bruni et al. 2018, p. 6). In contrast to the social sciences, the natural sciences have gone to great lengths to study the Kingdom Animalia, with its phyla, classes, orders, families, genus and species *per se* – their physiology, biological apparatuses and behaviours. They have been very successful in establishing themselves as capable of understanding and representing nonhuman animals. Within these natural science disciplines, anthropomorphism is considered an error to be avoided. However, anthropomorphism is often one of the main ways in which humans relate to animals and, thus, should also be seen as a varied set of specific HAR, which deserve scholarly investigation and reflection (Daston & Mitman, 2005; Parkinson. 2019; cf. also Servais, 2018) and which can be mobilised critically to explore human-animal entanglements (see Karlsson, 2012; Horowitz & Bekoff, 2007).

Relatedly, the fifth reason that some scholars voice incredulity towards HAS researchers interested in the animal side of HAR, is that often the pursuit of knowing "animals as such" (Shapiro 2020, p. 822; i.e.; what others have called "real animals" as opposed to representations of animals) is conflated with knowing animals as "pure" biological entities existing before the semiotic and symbolic orders that mediate our encounters. It is then important to clarify that we think it impossible to rid ourselves of all residual humanism and encounter "animals as such". This is because any human-animal relation – be it in the laboratory, under the microscope or in our everyday life – is always mediated by our (human) categories, biases and beliefs. The quest for knowing "what is like to be a bat", to recall Nagel's (1974) famous essay on consciousness and the limits of imagination, is thus impossible.

More modestly, from the perspectives of contemporary social sciences, HAS engages with "real animals" understood as material and semiotic presences and becomings; that is to say, the animal side of HAR we *can* explore is always at the same time both real and material, but also represented. It is fleshy, yet encountered through our very human social categories and perceptual apparatuses (our own human bodies). The animal side of HAR is not an absolute, but emerges relationally in specific material environments, which are always imbued with power (see Brown, this volume). In other words, across the social sciences, HAS investigates HAR by keeping in mind that these interactions are the product of a complex set of relations (socio-technical, cultural, economic, material, etc.). This kind of complexity is what perspectives from the social sciences offer and can add to the natural sciences to enrich current understandings of HAR (cf. Radhakrishna & Sengupta, 2020). Notably, the natural sciences have tended to disregard how the specificities of place, power, or gender, for example, shape how animals act in the laboratory (for a critique of the relations between objective knowledge, animals and the place of the experimental laboratory, see Despret, 2002, 2008). In summary, this book includes chapters about methods and methodologies to approximate and explore the animal side of HAR by keeping in mind that the "real" animals with which we can do research are always socio-semiotic and material presences.

We also argue that the sceptical reactions that scholars in HAS face stem from the fragmentation in discussions of methods – and thus perhaps the poor visibility of such discussions – and the never-ending call for innovative methods to deal with the animal side of HAR, which might leave the impression that HAS lacks methods. Next, as we note the inter- and transdisciplinary character of HAS, we point to previous work on methods, and we challenge the widespread assumption that (innovative) methods are lacking.

The long-standing and never-ending call for (innovative) methods

Today, HAS is a tremendously prolific and varied field. The inter- and transdisciplinarity of HAS contributes to nourishing its richness.[10] In 2010, DeMello identified 12 disciplines taking part in the scholarly debate. In 2020, Shapiro found 24 fields interfacing with HAS. As we write, we have counted 36 traditional and emerging fields and subfields that contribute to the study of HAR including, in alphabetical order: animal welfare science, anthropology, anthrozoology, archaeology, architecture, CAS, biosemiotics, cognitive ethology, comparative psychology, conservation biology, critical race studies, critical food studies, cultural studies, digital ecologies, disability studies, economic geography, feminist studies, geography, green criminology, history, law, literary criticism, media and digital studies, pedagogy, performance studies, philosophy, planning, political science, postcolonial studies, psychology, queer studies, sociology, science and technology studies, the visual arts, tourism studies, urban studies, veterinary humanities, ... and counting (see also Collado et al., 2022; Echeverri et al., 2018). These research fields do not share a singular method and methodology, but instead an interest in understanding animals and their relations with humans.

Despite this growing research with nonhuman animals, methods have rarely been discussed in detail. More precisely, existing discussions tend to be brief or scattered across diverse types of publication, such as research articles sections, full papers in specialised journals, single or groups of chapters within edited books.[11] As far as we are aware, only Hamilton and Taylor's *Ethnography after Humanism* (2017) focuses specifically on methods. Their book makes a strong case for extending a traditionally humanistic approach such as ethnography to research with nonhuman animals. It represents an important and accessible contribution as it spans a wide range of research methods drawn from HAR scholarship in the humanities and social sciences. This book extends Taylor and Hamilton's work by offering a compilation of social science research reflections, on ethnography as well as other methods, and focuses specifically on how our methods can better understand the animal side of our human-animal relationships.

Concept box 1.2: Multispecies ethnographies and animals

Multispecies ethnography is term now frequently used in reference to ethnography conducted with nonhumans, animals and otherwise, from a less anthropocentric perspective, notably introduced by Kirksey and Helmreich (2010). As the contributions in this book show, multispecies ethnographic approaches vary according to the notions used to design the methodology, and by the ways they explore the animal side of HAR. They can be generally considered as forms of doing and writing research that recognises the world as material and fluid, emerging from different ontologies and only partially knowable through research practice. Multispecies ethnographies involve engaging the world through an embodied attunement that pays attention to how the world itself, its component parts and forms of life are changing assemblages, ongoing compositions of connections and disconnections of multiple entities (see Ogden et al., 2013). These approaches tend to consider all beings and inorganic entities as surfacing through multiple ontologies, and emerging "via the present material configuration of multi-being connections" (Ogden et al., 2013, p. 15). Therefore, multispecies ethnographies complicate a fixed understanding of animals as material and semiotic "beings". Within these frameworks, animals are not simply "biocultural givens" (i.e., determined by biology and genes) and by "culture" (i.e., shaped by specific, geographic human beliefs and practices). Instead, animals are understood as "processes", rather than "beings": thus, animals "become with" (see Jones, this book, Chapter 4). Multispecies ethnographic approaches, then, "no longer reduce [animals and humans] to a species based on inherent capabilities (reproduction, for instance). [...] [They] *become* a particular kind of multispecies through their often precarious, unpredictable, and contingent relations with" (Ogden et al., 2013, p. 15; original emphasis) other beings, entities, forces, tensions, in specific environments and events shaped by human and nonhuman power. As they "move, incite, elicit and excite" humans and other beings, animals emerge as affective "naturecultures" (Latimer & Mara 2013, p. 8), assemblages of political, cultural, technological, environmental forces and entities (cf. Kirksey & Helmreich, 2010, p. 545).

More generally, multispecies (sometimes also called interspecies or cross-species) ethnographic approaches challenge traditional distinctions, underpinning modern Western ontologies, between humans and animals, culture and nature. They decentre human agency and offer ways for understanding how existence on this planet arises from the interconnections and disconnections of more-than-organic entities. Through conceptualising animals' agencies, subjectivities, capabilities, individualities, social lives and cultures, for example, multispecies ethnographic approaches offer a theoretical platform through which research methods can explore HAR from the perspectives of animals.

The fragmentation of the debate on methods is the outcome of intersecting trends in the field and, more generally, in academia. Besides the interdisciplinary nature of HAS, these trends include: the preservation and defence of disciplinary boundaries by research institutions also reflected in journals' aims and scopes; the politics of identity brought about by the different labels used to specify the approach for studying HAR (notably, anthrozoology, CAS, human-animal interaction studies, animal studies); and neoliberal, productivist trends which privilege the publication of articles and discourage the publication of more time-consuming books that delve in depth into questions of theory, practice and strategies – something this volume attempts to do.

The fragmented debate on methods matters as it has two interrelated main consequences. To begin with, the debate implies that HAS lacks solid methods – or might even need "*sui generis* methods" (Shapiro, 2020, pp. 823–824) – as the never-ending call for methods suggests by highlighting that, while the theorisation of HAR is advanced, methods are not (Margulies, 2019; see also Bell et al., 2018). With this book, we argue that while methods are crucial to the academic, practical and political pursuit of HAS, methods to explore the perspectives animals are not, in fact, lagging behind (see also Miele & Bear,

2023). As the richness of studies in HAS demonstrates, there is a plurality of methods which are flexible and can be adapted and combined to explore the animal side of HAR. The second consequence of such fragmentation, including the call for new methods, is that it could weaken HAS' influence and role in scholarly debates. Although the field appears to insiders to be established, the frequent reactions of scorn, derision or even anger towards HAR researchers in mainstream academia indicate that HAS might be "solid", yet still "at the margins", as Shapiro (2020) notably put it 20 years ago. The sense of inadequacy that researchers feel for doing "academic 'dirty work'" (Wilkie, 2015) might be, at least in part, counterbalanced by clarifying (ideally beyond the restricted circle of HAS scholars) what it means to emphasise the animal side of HAR in how we research with animals.

In short, the central argument in this book is that we can borrow and adapt tools from the social sciences to conduct research with nonhuman animals, combine them with knowledges and practices from the natural sciences, as well as collaborate with scholars and practitioners who work with animals. At our disposal, we already have a wide range of methods and research strategies that can be altered to deal with nonhuman research participants at all research stages, from research design and conceptualisation to field work and analysis. However, by arguing that we already have methods that can be adapted for research with animals, we do not claim to have finished our job of developing "innovative" (as the discourse of neoliberal academic parlance would have) tools. Rather, our argument is to make explicit that HAS is in position to be at the centre, and no longer at the margins, of critical and cutting-edge scholarship. So, how do we know animals? Or to be more precise how do we know *with* animals? Once the multiple possibilities of knowing with animals are clarified, as the chapters in this collection begin to do, then the field might be more acceptable, rather than an anathema. Below, we present an overview of how the book is organised and the central contributions of each chapter.

Theorising, collaborating and visualising human-animal relations: navigating the collection

The nine chapters that compose the book cover a broad spectrum of methods, which draw on various resources (material, textual, visual, multisensorial) and deploy different approaches (primarily, although not exclusively, multispecies ethnographic approaches) to conduct research with animals and interrogate them about their lives and experiences. The methodological diversity of contributions reflects the inter- and transdisciplinary character of HAS as the interventions are written by scholars and practitioners from diverse fields including sociology, geography, critical animal studies, feminist studies, environmental studies, wildlife conservation, ethology and anthropology. Each chapter appeals to both novice and more experienced researchers and practitioners in HAS through case studies and short concept boxes which walk the reader through common themes in HAS research. The book is organised into three broad thematic sections – "Theorising", "Collaborating" and "Visualising".

Theorising includes three chapters which emphasise that, in developing our research with animals, we need to adopt, adapt and deploy frameworks that enable our methods to focus on the experiences of animals. Erin Jones and Nik Taylor (Chapter 2 *Decentring humans in research methods: visibilising other animal realities*) posit that posthumanist inquiry should be informed by both ecofeminist and critical animal studies to develop methods which advocate for, and improve the lives of, the animals at stake in HAR relationships. To this end, they propose a method in which dog-owners participate in creative journaling about experiences of dog-walking, imagining the world from the perspectives of their pets. They compellingly argue that such a method actively prompts empathy for, and deeper attention to, dogs' experiences of the world and thus bring animals into the research as subjects, not just objects of study. While HAR are always mediated by our human interpretation of animal worlds, they nonetheless point out how a method like journaling can be adapted to foster perspective-taking and both confront and transcend the "limits" of anthropomorphism. Furthermore, Jones and Taylor iterate that posthumanism should be informed by advocacy and action, two central components of critical animal studies scholarship. Thus, they show how through methods which build empathy, humans change their behaviours towards their dogs and create more caring, and attentive, human-dog interactions. How to think of/with animals to better understand their points of view is further developed in the following two chapters which conceptualise, respectively, animals as intelligent subjects working with humans and as "becomings" who develop specific personalities.

Nicolas Lainé and Jocelyne Porcher (Chapter 3 *Understanding human-animal relations from the perspective of work*) offer an account of how farm animals in France and elephants in India communicate and work with their human handlers. This chapter proposes a framework for conceptualising animals as creative subjects who mobilise their intelligence and skills to accomplish tasks and establish diverse relations with humans and other animals. It provides a theoretical grid for HAR in the context of work which, according to the authors, has been humanity's principal form of life with nonhuman animals; i.e., the sphere that historically has tied humans and animals together, and where different species have come to know and shape each other. This contribution builds on Porcher's pioneering research on the work of animals (2002, 2017), which offers an alternative to Marxist theorisations of animals' exploitation and labour (cf. Barua, 2019; Wadiwel, 2018). Lainé and Porcher are sceptical of relying on Marxist thought because they argue it is difficult to see how animals might experience alienation through labour. Their methodological framework allows us to dive into the complexity of human-animal relations beyond a clear-cut matter of domination and exploitation (the two main relations that research on factory farming have brought to light), and to capture interspecies communication. This chapter introduces the reader to the importance of speaking with those who Despret (2008) calls "animals" good representers'; the humans who work with animals on an everyday basis, and who can speak to the capabilities which nonhuman creatures mobilise to accomplish working tasks.

Owain Jones's intervention (Chapter 4 *Re-thinking animal and human personhood: towards co-created narratives of affective, embodied, emplaced becomings of human and nonhuman life*) discusses two conceptual steps for thinking of HAR in a more symmetrical manner and focusing on animals' perspectives. First, he underscores the necessity of deconstructing the binary notions which frame humans as superior to other living creatures. Second, he argues it is central to draw on the ideas of affect and becoming, which help researchers focus on how humans and animals share embodied life, an existence lived and experienced primarily through the biological body, rather than reason and thought. These theoretical steps help us overcome concerns of anthropomorphism, focus on what animals and humans have in common (rather than on what makes them different) and move away from imagining animals as fixed entities situated "below" humans in a species hierarchy. More precisely, Jones conceptualises animals as individuals with biological repertoires who develop distinct personalities by living their lives in specific places with other individuals, humans and otherwise (cf. also Shapiro's, 1990, kinaesthetic empathy approach). To better understand the experiences of animals whose habitats, corporealities and biological endowments strikingly diverge from those of humans, he argues researchers can rely on insight from both academic experts and professional caregivers who study and work intimately with animals. This specific point is further developed in Part II of the book.

Collaborating discusses how social scientists can cooperate with practitioners who work closely with animals to better understand the complexities of our human-animal relationships. The section includes three chapters which present methods that connect research and practice. Here, methods are not only strategies to generate useful data for understanding HAR from the perspectives of other species, but also fundamental tools for including animals' needs into policy debates and initiatives. The methods articulated are thus explicitly and practically political as they aim to strengthen the recognition that society is more-than-human, that animals are its denizens who need to be recognised as stakeholders deserving respectful and inclusive policies. The first intervention in the second part of the book (Chapter 5 *Two species ethnography: honey bees as a case study of an interdisciplinary "more-than-human" method*) is penned by Siobhan Maderson and Emily Elsner-Adams, who develop a strategy to overcome two challenges in HAS: the investigation of (small) non-mammal species and the challenges of collaborative research with scholars from both social and natural sciences backgrounds. Building on their joint project on honey bees (*apis mellifera*), Maderson and Elsner-Adams propose "two-species ethnography" as an umbrella method attentive to the ways in which time and natural cycles shape fieldwork, and as a strategy combining tools and ideas from the social sciences, ethology and entomology. Two-species ethnography aims to generate data on the socio-economic and ecological contexts in which non-mammal species live. It builds on professionals' knowledge of their relations with animals, which helps "outsiders" such as social scientists get closer to the lives and experiences of those creatures, whose bodies, habitats and Umwelt strikingly differ from those to which humans are accustomed. Two-species ethnography thus serves as a guide to support research by offering a holistic and ethologically-informed understanding of the lives of non-mammals, so that we can better attend to what they need and avoid practices which could harm them.

Hannah Jaicks-Ollenburger's piece (Chapter 6 *Trekking a predator's journey: paths through the Greater Yellowstone Ecosystem*) focuses on the collaboration between researchers and wildlife managers to re-visit conservation policies' traditional emphasis on humans and their dismissal of animals' needs. The chapter recounts the author's fieldwork following and experiencing the migration routes of grizzlies, cougars and wolves across Mexico, the United States, and Canada. Theorising from a wide spectrum of intellectual resources (e.g., feminist science studies, phenomenology and multispecies ethnography), Jaicks-Ollenburger develops "trekking" as a practical, mobile method for generating data on how animals experience life in an increasingly humanised nature.[12] While predators are frequently framed as animals dangerous to humans, in Jaicks-Ollenburger's framework, the roles are inverted: it is humans' activities who pose dangers to predators. In decentring safety, this methodological apparatus is particularly useful to bring to light the many, human-made obstacles that animals encounter in everyday life. Trekking thus pinpoints the various human-animal conflicts that arise with the increasing encroachment of humans into previously "wild" spaces.

Susan Boonman-Berson and Séverine van Bommel's Chapter 7 (Chapter 7 *How to do multispecies-ethnographies when exploring human-(wild) animal interactions: affect, multisensory communication and materiality*) highlights the need to understand how science and policy should change their practices of managing wildlife *with* animals and not against them (i.e., culling). The authors advocate for conducting research across the social and natural sciences by adopting methods and knowledges from ethology and ethnography and by collaborating with practitioners. In Boonman-Berson and van Bommel's research, wildlife managers are understood as "interpreters" of animals, as they are equipped with the skills and the capacities to communicate with animals by mobilising multisensorial linguistic codes acquired through extensive contact with animals. Drawing on their work with elephants in Kenya, black bears in Colorado and wild boars in the Netherlands, the authors discuss how to put into practice multispecies ethnographies that consider animals as stakeholders, and that specifically centre on the affective and sensorial registers of interspecies communication. They argue that understanding and translating how animals "talk" to each other, including to humans, is imperative as we design conservation strategies attentive to the needs of animals.

The emphasis on the senses for exploring HAR is further elaborated in the third and final section of the volume, *Visualising*, which presents how visual methods can produce useful data of HAR by focusing on our somatic and embodied registers. While visual methodologies have become common in the social sciences and humanities in the last 20 years or so,[13] human-animal scholars have only recently adopted a variety of visual strategies (see Hamilton & Taylor, 2017, pp. 89–109). In Chapter 8 (*Shared sensory signs: mine detection rats and their handlers in Cambodia*), Darcie DeAngelo recounts her fieldwork in the Cambodian minefields to better understand how human and rat deminers work together. DeAngelo shows how filming is an effective method for generating data on interspecies communication and miscommunication. To theorise, and analyse, interspecies communication, she draws on Peirce's (1998) and Kohn's (2013) semiotics and on the idea that the perspective of rats can be investigated because their bodies are equipped with the same senses that humans possess – although these are differently developed and used. Scholarship in ethology and biology offers crucial insight for understanding how specific species experience and perceive the world, communicating with others with somatic apparatuses in ways which are fundamentally different from humans. In posing a simple question about love between rats and their handlers, DeAngelo asks, do animals reciprocate this feeling? While human co-workers claim to feel love for the Gambian Pouched rats they train and handle, what can we know about the rats' perspectives? The video camera becomes her tool for producing visual data and for decentring and, in some cases, challenging the perspectives of humans. DeAngelo's ethnography exposes tensions between human-centred narratives and the intimate portrayal of subtle, quick moments and overlooked action which emerge through her analysis of film recordings. Experimental, creative writing is an additional method she uses to illuminate individual rats' point of view and thus reveal humans' perspectives on the animals with whom they interact and work.[14] It is also a useful tool for emphasising the ethical dilemmas which arise when human-animal scholars and practitioners work with other creatures who can be harmed, endangered or neglected.[15] Importantly, writing is understood as a performative research strategy that intervenes in, rather than simply mirrors, reality itself. The potential of methods to intervene in, and possibly change, the world is particularly important for human-animal scholars and practitioners interested in improving human-animal relations (see also Jones & Taylor, Chapter 2). This ability of methods to affect how humans, animals and the environment interrelate is reiterated in the final two chapters, as well.

In Chapter 9 (*Doing multispecies ethnography with mobile video: exploring human-animal contact zones*), Katrina Brown elaborates on the potential, and the limits, of videorecording techniques to understand cross-species and embodied communication, and the ethical dilemmas which she confronts. Her methodology is based on notions such as becoming-with, affect, embodiment and movement (see Jones' Chapter 4 for a discussion of these concepts) and she shows how these ideas can be applied when analysing data produced through mobile-video ethnography with wearable video cameras. These devices create visualisations of how embodied life comes into being through human-animal encounters, emerging in distinct ways, as bodies move through space. Building on Haraway's idea of "contact zones" (2010, 2008), Brown discusses her mobile video ethnography of dog-walking in Cairngorms National Park, UK. Similarly to DeAngelo, Brown argues that doing research with video cameras requires reflexivity on how animals and their relations with humans and their environment are made visible, invisible, audible or imperceptible through filming. Far from capturing an absolute reality, the film thus produced offers the opportunity to reflect on the sensorial hierarchies that researchers create as they perceive the audio and the visuals during data analysis. Whilst DeAngelo focuses on quick movements and close-up perspectives, Brown comments instead on how stillness and silences captured in the video footage are analytical cues for delving into animals' agencies, their interests, and interactions with specific places and other animals, human and otherwise.

In the last contribution of this book, Chapter 10 (*Getting visceral: body mapping the humanimalian*), Heide K. Bruckner tackles another of the most mundane sites of human encounters with animals, where their lives and bodies are, perhaps paradoxically, mostly rendered imperceptible: the eating of meat. Eating animal products is an ordinary contact zone where human-animal relations are articulated yet taken for granted and thus made invisible. How do we explore the taken for granted-ness of animals' presences in things and practices which appear to be unrelated to their animal origins? Bruckner discusses how body mapping – an arts-based research method to produce data and to communicate findings to a wider audience (see Aroussi et al., 2022) – can make ordinary HAR visible and palpable again. Her methodology pulls from feminist scholarship on the body as porous and more-than-human; that is to say, the body is understood as a place which is open and permeable, as a process rather than a fixed container of a discrete human self, separated from apparently external socio-economic formations and the environment. It also builds of the notion of the visceral; namely, those sensations bodies feel in a pre-conscious state and express without the aid of words, but with gestures and somatic expressions. The chapter outlines the practicalities of how to conduct body mapping research, from recruiting participants to the analysis of the data. In body mapping (but also other visual methods), images that portray animals and human-animal relations can be used to evoke specific, affective responses in their viewers and lead to reflection on taken-for-granted practices that involve, yet conceal or degrade, animals. In so doing, visual research strategies act as methods for intervening in the world and challenging current anthropocentric structures, discourses and practices.

A common thread through the diverse contributions of the book is the critical importance of methods as performative – i.e., crucial tools that help make realities. To understand methods in this way, Law and Urry (2004) argue, means being upfront with yourself as a researcher as to which "realities" you seek to make more "real". The responsibility becomes more than applying a technical tool to the world; instead, it requires deeper inquiry into the politics and polemics of the methods at hand. Our interest in the theorising, collaborating and visualising threads throughout the book are informed by what we see as the three central contributions of the book. First, we aim to engage with the controversies surrounding HAS and its pursuit of delving into the animal side of HAR. Second, we want to present a diversity of practical tools about how we can better understand animals' individual and collective experiences. Clarifying what scholars interested in the animal side of HAR aim to do, and how, is crucial for bringing HAS from the margins to the centre of current academic, professional, and policymaking endeavours. Third, we emphasise that methods are crucial for HAS' pursuit of reconnecting humans and animal worlds, and we hope that by developing the methodological tools to pay attention and become attuned to animals' lives, we can co-create more-than-human socioecological arenas where humans and animals coexist in less deadly ways.

Despite HAS being a varied, well-established, and prolific field, there is still plenty of work to be done to find ways to reconnect and bring to the forefront the needs of other-than-human creatures. In the conclusion, we point to the limits of the present volume, and we suggest future scenarios and ways forward for methods that explore animals' perspectives.

Conclusion

While the chapters vary in approach and case study, ultimately this book primarily focuses on variations of multispecies ethnography and on terrestrial mammals. Furthermore, it does not include strategies for investigating HAR in cyberspace, nor does it include quantitative methods.

The reliance on multispecies ethnography (see Concept box 1.2, this chapter) echoes the current, dominant approach in the social sciences to do research with animals. Future methods, within and beyond multispecies ethnographic approaches, could be developed by drawing on other-than-Western ontologies and knowledges (see Watson & Huntington, 2008). Researchers, including the authors of this chapter, in "international" (English-speaking) academia are starting to recognise their own positionality within the Minority World. Scholars, theories, knowledges and expertise from the Majority World remain largely unrepresented in academia and policymaking. This bias is reflected in current HAS, although so-called "indigenous studies" and knowledges are gaining prominence. Because of structural problems such as global socio-economic inequalities and local power relations in academia – not to mention the widespread idea that animals are either irrelevant or impossible to understand – the "globalisation and decolonisation" (Hovorka, 2017) of HAS is a project which requires time and resources, and which should be further pursued. This can be done by exploring how the symbolic and material becomings of specific animals in different places beyond the Minority World foster diverse human-animal relations and, relatedly, different animals' perspectives and experiences. We need to know more about animals as material-semiotic achievements (simultaneously, creatures in the flesh and socially constructed) in different geographical contexts and cultures, and from other-than-Western theoretical perspectives and ontologies for at least two main reasons. First, it allows us to further investigate how the animals in us have been hidden by diverse anthropogenetic machines.[16] We can then develop ways of empathic reconnection and senses of reciprocity for putting into practice less exploitative human-animal relations. Relatedly, doing research with animals in diverse geographies in the Majority World by either mixing our theoretical toolkits or by adopting entirely different ontological frameworks represents an important step to enrich our knowledge of the wide spectrum of HAR animating the planet. Novel multispecies, and multi-sited, ethnographic approaches expanding mobile methods, for example, to follow the lives and deaths of animals in diverse geographical contexts could thus provide useful insights on which practices and relations should be avoided or adopted/adapted to benefit animals in policy measures and to pursue less conflictual engagements between species.

With the exception of Maderson and Elsner Adams' two-species ethnographical platform to study honey bees, the chapters in this book present tales of research of "minority species"; i.e., terrestrial mammals. The book thus overlooks other creatures who live underground, in the water and/or across the sky. This limitation is another major trend in current HAS in the social sciences; namely, the myopic focus on mammals. It also asserts the need for thinking about methodologies able to delve into other animals' worlds to investigate the points of views of other-than-mammal creatures and more-than-mammal HAR (see e.g., Azevedo et al., 2022, on humans' bonds with reptiles and Shipley & Bixler, 2017, on humans' understandings of arthropods and Bear, 2019, analysis of insect farming practices). These present humans with a range of challenges which are not easy to address. Birds, amphibians, reptiles, insects, fish and other creatures have bodies which are strikingly different from human ones. Or, they are invisible to us because of their microscopic sizes or are difficult to perceive as they live in and across environments that are not easily accessible to humans, as we lack specific somatic apparatuses and capabilities to experience most of the richness of the worlds across which other animals live. More-than-mammalian animals, diversely classified in millions of species, live across a plurality of worlds and human and animal-made infrastructures (termite mounds, anthills and beaver dams, for example). Researchers in the social sciences are presented with real challenges when aiming to understand the animal side of HAR in contexts characterised by different depth of the soils and waters, diverse altitudes, landscapes and environments less habitable for humans. How do we explore such contexts which bring together different worlds and experiences from other non-mammals' points of views? To tackle these challenges, HAS could adopt pragmatic (as opposed to dogmatic) approaches, as suggested by this book's part on collaboration, and put into practice cooperative plans and strategic platforms that bring practitioners together with scholars in the natural and social sciences, but also leverage the expertise of volunteers, advocates and others who work and live closely with animals.[17] Although academic curricula preparing future HAR experts across discipline diverse are being developed, primarily in the Minority World,[18] it is likely that, in the near future, research will entail collaborations across expertise, academic and otherwise. Perhaps such cooperation and institutional arrangements will generate novel and specific methods for studying the animal side of HAR.[19]

Social media, virtual and augmented realities and surveillance technologies used to monitor animals on earth increasingly produce and re-produce novel and old human-animal encounters. Scholars in HAS in diverse disciplines, including the emerging field of digital ecologies (Searle et al., 2022),[20] are mobilising digital methodologies to explore HAR in new spaces and infrastructures. In these internet mediated contexts, methods appear to be more useful to delve into the human side of HAR by analysing how humans (dis-)entangle with animals or negotiate cross-species relations or by examining how humans control animal individuals and populations (see Berland, 2019; Brower, 2021; von Essen et al., 2021). However, mixed methods that consider both digitised animals in cyberspace and their fleshy counterparts as studied by biology and ethology, for example, could be adapted to delve into the animal side of HAR. Novel findings in the natural sciences, including the discovery of other species, more precise monitoring technologies and the internet do multiply human-animal encounters that need to be considered. Methods could be adapted or developed to follow such proliferation and bring to light other ways to think about how we foster multispecies convivialities.

Despite the increasing visibility of animals (in the media, cities and our homes), the growing numbers of animals produced and killed in factory farming to sustain rising demand for meat, or the acceleration of the sixth mass extinction, numbers do not seem to move human societies to change the ways we treat animals. Yet, policy measures are often influenced by statistics and figures. While from the social sciences' perspective the mobilisation of quantitative methods to explore the animal side of HAR appears difficult, collaboration across diverse disciplines and quantitative methods might assist (rather than prevent) in making animals and their needs appear more urgent and politically relevant. First, the wide spectrum of HAR can be measured and quantified (see Echeverri et al., 2018; Possidónio et al. 2019). Second, developments in technologies for monitoring animals' mobilities, advancements in the natural sciences in the study of the mind and bodies of animals, and the evolution of machine learning tools might offer future opportunities for generating data by identifying, estimating and measuring a wide of emotions and other visceral, cognitive and physiological processes at stake in both sides of HAR.

Frequent episodes that involve unnecessary killings of animals, for example, culling of farm animals during Covid 19 or the recent high-profile killing of the walrus Freya,[21] show how we still live in societies which are profoundly anthropocentric and in which animals' lives are easily expendable. Despite HAS being a prolific field of study, conducting research with animals in the social sciences, in our experience, is still perceived either as irrelevant, unnecessary or futile and ridiculous. We hope this book encourages HAS students and scholars to be more assertive and less defensive, and further engage in methodologies that highlight how the animal-side of human-animal relations can be addressed through research that is always partial, positioned and performative. Such knowledge might be "humble": it neither has the presumption of mirroring the reality nor inner worlds of animals. It should, however, hold the potential to teach us how to behave more carefully and responsibly with nonhuman animals, an imperative in the age of Anthropocene.

Notes

1 These anonymised opening quotes are representative of some of the many remarks we have overheard in the last ten years or so while speaking with our social scientist colleagues about our research interests at work, conferences and during proposal planning meetings in continental Europe. While it might be true that animals are conceptualised through instrumentalist and utilitarian lenses primarily in STEM (science, technology, engineering and math) disciplines, here we wish to emphasise that in our experience, ideas that animals "do not matter" and are fundamentally inferior – and thus ignorable and invisible – are rather widespread also in the social sciences, perhaps more prominently in academic contexts characterised by very strict disciplinary boundaries that hinder trans- and interdisciplinarity, which are at the very core of HAS. For a discussion of how human-animal scholars may be tainted, see Wilkie (2015).

2 See Deem et al. (2019) for a discussion of the One Health approach; Seltenrich (2018) for the Planetary Alliance program.

3 See Sebo (2022) for an excellent discussion on how humans' treatment of animals matters for climate change, biodiversity loss and other disasters characterising the Anthropocene.

4 For an up-to-date list of journals, courses and centres, refer to the Animal & Society Institute's excellent website https://www.animalsandsociety.org/resources/resources-for-scholars/ (last accessed 10/08/2022).

5 See Jones and Taylor, this volume, and Collado et al.'s (2022) bibliometric analysis for further discussions of the terminology animating the field.

6 We adopt Alam's (2008) terminology "Minority World" and "Majority World", instead of Global North and Global South or Developed and Developing World, because the latter are expressions that incorporate profound biases and assumptions about the ways in which the world should "develop", hide other than neoliberal and Western ways of being, doing and thinking. The Minority World comprises the wealthiest areas of the planet where mainly white, relatively educated people live primarily in cities and which represent just a small part of the world population. The Majority World "is where most people live: '[I]t…' covers Asia, Africa, the Arabic-speaking world, Latin America, the Caribbean, and the Indigenous Peoples of the planet" (Tandon & Hall 2014, p. 54, cit. in Kneafsey et al. 2021, p. 74).

7 See Dunlap and Mertig (2014) on the history of American environmental movements; Best (2009) and DeMello (2021, pp. 470–498), respectively, on the animal liberation and animal protection movements.

8 See endnote 1.

9 Besides the limits of language in capturing "reality", conceptualisations of radical otherness, stemming from the work of Levinas (1969) and Derrida (2008) also point to how we are never able to fully know each other, including nonhuman animals.

10 For books that offer overviews of HAS, see DeMello (2021); Kalof (2017); Marvin and McHugh (2014); Roscher et al. (2021); Hollin (2021).

11 See Bastian, 2016; Hovorka et al., 2021; Taylor & Twine, 2015. For example, *The Rise of Critical Animal Studies* (Taylor & Twine, 2015) includes a part with three chapters dedicated to "Doing Critical Animal Studies", which focuses on mixed methods (tracing networks, ethnographic observations, biographical and visual research strategies) for investigating animals' lives and presences, and explores how power shapes human-animal interactions in specific contexts (Birke, 2014; Gröling, 2014; Griffin, 2014). Alice Hovorka et al.'s (2021) edited volume on advancing the scope of animals' geographies also has a part dedicated to the exploration of animals' "Lifeworlds", where the authors discuss multispecies ethnographic approaches, walking methods and affective ethology (Ellis, 2021; Arathoon, 2021; MacKay, 2021; Sinha et al. 2021). Bastian et al. (2016) have edited an inspiring book on participatory research methods with nonhuman agents. It also includes interventions on thinking about animals as research partners in ethnographic explorations, developing technologies which can be used by nonhuman animals, and on theatre making in order to find ways to bring to light animals' agencies, give voice and listen to the needs of other than human animal creatures (Bastian, 2016; Taylor, 2016; Mancini, 2016; Hodgetts, 2016; Heddon, 2016).

12 On mobile methods in HAS, see Arathoon (2021).

13 For a useful historical overview of how visual data and visual methods have been integrated in social science research, see Watt and Wakefield (2017).

14 On creative writing as a research method, see Cook (2013) and Crewe (2021).

15 Doing research with animals is always an ethical pursuit, and ethical dilemmas are likely to arise while doing fieldwork. Work on ethics and animals is far from lacking (e.g., Fischer, 2019; Gruen, 2021), and there are official guidelines for conducting research on animals in the natural sciences. Yet, as far as we are aware, in the social sciences there are no official guidelines for doing research with animals. As HAS is becoming increasingly relevant in academia, and funding bodies are beginning to support projects that explore HAR within the social sciences and humanities, researchers in the social sciences need official guidelines. The lack of official guidelines for ethical conduct in HAS shows, once more, how animals might be irrelevant (cf. Oliver, 2021). There is an institutional gap that needs to be filled, as discussions of the ethics of doing research with animals are critical for the intellectual and political pursuit of ameliorating the lives of animals.

16 The anthropogenetic (or anthropological) machine is a notion made popular and developed by Agamben (2004) and can be understood as the visual, cultural institutional device that has historically separated (some) humans from animals in the West. Such device, which keeps operating, can be described as the ensemble of the institutions and related narratives that work together to keep separate the animal registers of humans and reproduce the myth of human supremacy.

17 See Echeverri et al. (2018, pp. 56–57) on cross-paradigmatic research.

18 For an up-to-date list, see https://www.animalsandsociety.org/resources/resources-for-students/degree-programs/

19 In the most recent review of HAS, Shapiro (2020) asks if the field will need "*sui generis* methods" to study the "animals as such", rather than keep adapting those we have at our disposal. Far from excluding the possibility of developing novel methods, we nonetheless think that imagining that we need specific methods to explore animals might reiterate the impression of the abyssal difference of animals and their related inscrutability.

20 On the emerging field of digital ecologies, see http://www.digicologies.com/about/ (last accessed 25 August 2022).

21 On the culling of farm animals during the Covid 19 pandemic, see e.g. https://www.theguardian.com/environment/2020/may/19/millions-of-us-farm-animals-to-be-culled-by-suffocation-drowning-and-shooting-coronavirus (last accessed 29 August, 2022). On the likely unnecessary killing of the Freya because humans could not be convinced to keep a reasonable distance from the walrus, see https://www.bbc.com/news/world-europe-62556295 (last accessed 3 September 2022).

References

Agamben, G. (2004) *The Open: Man and Animal*. Stanford, CA: Stanford UP.

Alam, S. (2008) "Majority world: Challenging the West's rhetoric of democracy." *Amerasia Journal, 34*(1), pp. 88–98.

Arathoon, J. (2021) "Researching animal geographies through the use of walking methods." In Hovorka, A., McCubbin, S. and Van Patter, L. (eds.) *A Research Agenda for Animal Geographies*. Northampton: Edward Elgar Publishing, pp. 101–114.

Aroussi, S., Badurdeen, F.A., Verhoest, X. and Jakala, M. (2022) "Seeing bodies in social sciences research: Body mapping and violent extremism in Kenya." *Qualitative Research*. https://doi.org/10.1177/14687941221096598

Azevedo, A., Guimarães, L., Ferraz, J., Whiting, M. and Magalhães-Sant'Ana, M. (2022) "Understanding the human–reptile bond: An exploratory mixed-methods study." *Anthrozoös*, pp. 1–18. https://doi.org/10.1080/08927936.2022.2051934

Barua, M. (2019) "Animating capital: Work, commodities, circulation." *Progress in Human Geography, 43*(4), pp. 650–669.

Bastian, M. (2016) "Towards a more-than-human participatory research." In Bastian, M., Jones, O., Moore, N. and Roe, E. (eds.) *Participatory Research in More-Than-Human Worlds*. London: Routledge, pp. 33–51.

Bear, C. (2019) "Approaching insect death: Understandings and practices of the UK's edible insect farmers." *Society & Animals, 27*(7), pp. 751–768.

Bell, S.J., Instone, L. and Mee, K.J. (2018) "Engaged witnessing: Researching with the more-than-human". *Area, 50*(1), pp. 136–144.

Berger, J. (1980) "Why look at animals?" In *About Looking*. New York: Pantheon.

Berland, J. (2019) *Virtual Menageries: Animals as Mediators in Network Cultures*. Cambridge, MA: MIT Press.

Best, S. (2009) "Rethinking revolution: Total liberation, alliance politics, and a prolegomena to resistance movements in the twenty-first century." In *Contemporary Anarchist Studies*. London: Routledge, pp. 205–215.

Birke, L. (2014) "Listening to voices. On the pleasures and problems of studying human-animal relationships." In Taylor, N. and Twine, R. (eds.) *The Rise of Critical Animal Studies: From the Margins to the Centre*. London: Routledge, pp. 71–87.

Braidotti, R., & Hlavajova, M. (eds.) (2018) *Posthuman Glossary*. London: Bloomsbury Publishing.

Brower, M. (ed.) (2021) *Exhibiting Digital Animalities*. Toronto: PUBLIC Books.

Bruni, Domenica, Pietro Perconti, and Alessio Plebe. (2018) "Anti-anthropomorphism and its limits." *Frontiers in Psychology, 9*, pp. 2205.

Buller, H. (2015) "Animal geographies II: Methods." *Progress in Human Geography, 39*(3), pp. 374–384.

Calarco, M. (2008) *Zoographies: The Question of the Animal from Heidegger to Derrida*. New York: Columbia UP.

Calarco, M. (2020) *Animal Studies. The Key Concepts*. London: Routledge.

Calvo, E. (2008) "'Most farmers prefer blondes': The dynamics of anthroparchy in animals' becoming meat." *Journal for Critical Animal Studies, 7*(1), pp. 32–45.

Carlson, C.J., Albery, G.F., Merow, C. et al. (2022) "Climate change increases cross-species viral transmission risk." *Nature, 607*, pp. 555–562. https://doi.org/10.1038/s41586-022-04788-w

Cartmill, M. (2000) "Animal consciousness: Some philosophical, methodological, and evolutionary problems." *Behavioural Processes, 52*, pp. 89–95. https://doi.org/10.1093/icb/40.6.835

Collado, E.B., Martín, P.T. and Serena, O.C. (2022) "Mapping human-animal inter-action studies: A bibliometric analysis." *Anthrozoös*. https://doi.org/10.1080/08927 936.2022.2084994

Colombino, A. and Giaccaria, P. (2021) "The Posthuman imperative: From the question of the animal to the questions of the animals." In Tanca, M. and Tambassi, M. (eds.) *The Philosophy of Geography*. Cham: Springer, pp. 191–210.

Cook, J. (2013) "Creative writing as a research method." In Griffin, G. (ed.) *Research Methods for English Studies*. Edinburgh: Edinburgh UP, pp. 200–217.

Crewe, J. (2021) "Creative writing as a research methodology." *New Vistas, 7*(2), pp. 26–30.

DeMello, M. (2021) *Animals and Society: An Introduction to Human-Animal Studies* (2nd ed.). Columbia University Press. https://doi.org/10.7312/deme19484

Daston, L. and Mitman, G. (eds.) (2005) *Thinking with Animals: New Perspectives on Anthropomorphism*. New York: Columbia University Press.

Deckha, M., (2012) "Toward a postcolonial, posthumanist feminist theory: Centralizing race and culture in feminist work on nonhuman animals." *Hypatia, 27*(3), pp. 527–545.

Deem, S.L., Lane-deGraaf, K.E. and Rayhel, E.A. (2019) *Introduction to One Health: An Interdisciplinary Approach to Planetary Health*. Hoboken, NJ: John Wiley & Sons.

Derrida, J. (2002) "The animal that therefore I am (more to follow)," trans. David Wills, *Critical Inquiry, 28*, pp. 369–418.

Derrida, J. (2008) *The Animal that Therefore I Am*. New York: Fordham University Press.

Despret, V. (2002) *Quand le loup habitera avec l'agneau*. Paris: les Empêcheurs de Penser en rond.

Despret, V. (2008) "The becomings of subjectivity in animal worlds." *Subjectivity, 23*(1), pp. 123–139.

Dunlap, R.E. and Mertig, A.G. (2014) *American Environmentalism: The US Environmental Movement, 1970–1990*. London: Taylor & Francis.

Echeverri, A., Karp, D.S., Naidoo, R., Zhao, J. and Chan, K.M. (2018) "Approaching human-animal relationships from multiple angles: A synthetic perspective." *Biological Conservation, 224*, pp. 50–62.

Ellis, R. (2021) "Sensuous and spatial multispecies ethnography as a vehicle to the re-enchantment of everyday life: A case study of knowing bees." In Hovorka, A., McCubbin, S. and Van Patter, L. (eds.) *A Research Agenda for Animal Geographies A Research Agenda for Animal Geographies*. Northampton: Edward Elgar Publishing, pp. 87–100.

Fischer, B. (ed.) (2019) *The Routledge Handbook of Animal Ethics*. London: Routledge.

Fraiman, S. (2012) "Pussy panic versus liking animals: Tracking gender in animal studies." *Critical Inquiry, 39*(1), pp. 89–115.

Gaard, G. (1993) *Ecofeminism*. Philadelphia: Temple University Press.

Gibbons, M., Sajedeh, S. and Lars, C. (2022) "Descending control of nociception in insects?" *Proceedings of the Royal Society B: Biological Sciences*. https://doi.org/10.1098/rspb.2022.0599

Griffin, N.S. (2014) "Doing critical animal studies differently." In Taylor, N. and Twine, R. (eds.) *The Rise of Critical Animal Studies: From the Margins to the Centre*. London: Routledge, pp. 111–136.

Gröling, J. (2014) "Studying perpetrators of socially-sanctioned violence against animals through the I/eye of the CAS scholar." In Taylor, N. and Twine, R. (eds.) *The Rise of Critical Animal Studies: From the Margins to the Centre*. London: Routledge, pp. 88–110.

Gruen, L. (2021) *Ethics and Animals: An Introduction*. Cambridge, UK: Cambridge University Press.

Hamilton, L. and Taylor, N. (2017) *Ethnography after Humanism: Power, Politics and Method in Multi-species Research*. London: Springer.

Haraway, D. (1988) "Situated knowledges: The science question in feminism and the privilege of partial perspective", *Feminist Studies*, *14*(3), pp. 575–599.

Haraway, D.J. (2008) *When Species Meet*. Minneapolis: University of Minnesota Press.

Haraway, D.J. (2010) "Training in the contact zone." In Da Costa, B. and Philip, K. (eds.) *Tactical Biopolitics. Art, Activism, and Technoscience*. Cambridge, MA: MIT Press, pp. 445–464.

Harrison, R. (1964) *Animal Machines*. London: Vincent Stuart.

Heddon, D. (2016) "Con-versing: Listening, speaking, turning." In Bastian, M., Jones, O., Moore, N. and Roe, E. (eds.) *Participatory Research in More-Than-Human Worlds*. London: Routledge, pp. 206–222.

Hodgetts, T. (2016) "How we nose." In Bastian, M., Jones, O., Moore, N. and Roe, E. (eds.) *Participatory Research in More-Than-Human Worlds*. London: Routledge, pp. 93–105.

Hollin, C.R. (2021) *An Introduction to Human–Animal Relationships: A Psychological Perspective*. New York: Routledge.

Horowitz, A.C. and Bekoff, M. (2007) "Naturalizing anthropomorphism: Behavioral prompts to our humanizing of animals." *Anthrozoös*, *20*((1), pp. 23–35.

Hovorka, A. (2017) "Animal geographies I: Globalizing and decolonizing." *Progress in Human Geography*, *41*(3), pp. 382–394.

Hovorka, A., McCubbin, S. and Van Patter, L. (eds.) (2021) *A Research Agenda for Animal Geographies*. Northampton: Edward Elgar Publishing.

Kalof, L. (ed.) (2017) *The Oxford Handbook of Animal Studies*. New York: Oxford UP.

Karlsson, F. (2012) "Critical anthropomorphism and animal ethics." *Journal of Agricultural and Environmental Ethics*, *25*, pp. 707–720.

Kirksey, S.E. and Helmreich, S. (2010) "The emergence of multispecies ethnography." *Cultural Anthropology*, *25*(4), pp. 545–576.

Kneafsey, M., Maye, D., Holloway, L. and Goodman, M.K. (2021) *Geographies of Food: An Introduction*. London: Bloomsbury Publishing.

Kohn, E. (2013) *How Forests Think Toward an Anthropology beyond the Human*. Berkeley: Univ of California Press.

Latimer, J. and Mara, M (2013) "Naturecultures? Science, affect and the non-human." *Theory, Culture & Society*, *30*(7), pp. 5–31.

Latour, B. (2005) *Reassembling the Social: An Introduction to Actor-Network-Theory*. Oxford: Oxford University Press.

Law, J. and Urry, J. (2004) "Enacting the social." *Economy and Society*, *33*(3), pp. 390–410.

Levinas, E. (1969) *Totality and Infinity*. Pittsburgh: Duquesne.

MacKay, C. (2021) "Animal subjectivities and lifeworlds: Working with and learning from animals through the practice of multispecies participant observation." In Hovorka, A., McCubbin, S. and Van Patter, L. (eds.) *A Research Agenda for Animal*. Northampton: Edward Elgar Publishing, pp. 115–128.

Magouras, I., Brookes, V.J., Jori, F., Martin, A., Pfeiffer, D.U. and Dürr, S. (2020) "Emerging zoonotic diseases: Should we rethink the animal–human interface?" *Frontiers in Veterinary Science*, *7*. https://doi.org/10.3389/fvets.2020.582743

Mancini, C. (2016) "Animal–computer interaction: A manifesto." In Bastian, M., Jones, O., Moore, N. and Roe, E. (eds.) *Participatory Research in More-than-Human Worlds*. London: Routledge, pp. 68–79.

Margulies, J.D. (2019) "On coming into animal presence with photovoice." *Environment and Planning E: Nature and Space*, *2*(4), pp. 850–873.

Marvin, G. and McHugh, S. (eds.) (2014) *Routledge Handbook of Human-Animal Studies*. London: Routledge.

Miele, M. and Bear, C. (2022) "Geography and posthumanism." In Herbrechter, S., Callus, I., Rossini, M., Grech, M., de Bruin-Molé, M. and Müller, C.J. (eds.) *Palgrave Handbook of Critical Posthumanism*. Cham: Springer International Publishing, pp. 1–23. https://doi.org/10.1007/978-3-030-42681-1

Miele, M. and Bear, C. (2023) "More-than-human research methodologies." in Clifford, N. Cope, M., Gillespie, T. and French, S. (eds.) *Key Methods in Geography*. Thousand Oaks, CA: Sage.

Nagel, T (1974) "What is it like to be a bat?" *The Philosophical Review*, *83*(4), pp. 435–450.

Nibert, D. (2013) *Animal Oppression and Human Violence: Domesecration, Capitalism, and Global Conflict*. New York: Columbia University Press.

Ogden, L.A., Hall, B. and Tanita, K. (2013) "Animals, plants, people, and things: A review of multispecies ethnography." *Environment and Society*, *4*(1), pp. 5–24.

Oliver, C. (2021) "Beyond-human ethics: The animal question in institutional ethical reviews." *Area*, *53*(4), pp. 619–626.

Pacini-Ketchabaw, V., Taylor, A. and Blaise, M. (2016) "Decentring the human in multispecies ethnographies." In Taylor C. and Hughes, C. (eds.) *Posthuman Research Practices in Education*. London: Palgrave Macmillan, pp. 149–167.

Parkinson, C. (2019) *Animals, Anthropomorphism and Mediated Encounters*. London: Routledge.

Pedersen, H. (2011) "Release the moths: Critical animal studies and the posthumanist impulse." *Culture, Theory and Critique*, *52*(1), pp. 65–81.

Peirce, C.S. (1998) *The Essential Peirce: Selected Philosophical Writings*, vol. 2 (1893–1913) edited by the Peirce Edition Project. Indiana UP: Bloomington.

Porcher, J. (2002) *Eleveurs et animaux: réinventer le lien*. Paris: PUF.

Porcher, J. (2017) *The Ethics of Animal Labor: A Collaborative Utopia*. Cham: Springer.

Possidónio, C., Graça, J., Piazza, J., & Prada, M. (2019) "Animal images database: Validation of 120 images for human-animal studies." *Animals*, *9*(8), p. 475.

Quammen, D. (2012) *Spillover: Animal Infection and the Next Human Pandemic*. New York: W.W. Norton & Company.

Radhakrishna, S. and Sengupta, A. (2020) What does human-animal studies have to offer ethology? *Acta ethologica*, *23*(3), pp. 193–199.

Roscher, M., Krebber, A., & Mizelle, B. (eds.). (2021) *Handbook of Historical Animal Studies*. München: Walter de Gruyter GmbH & Co KG.

Rose, J.D., Arlinghaus, R., Cooke, S.J., Diggles, B.K., Sawynok, W., Stevens, E.D. and Wynne, C.D. (2014) "Can fish really feel pain?" *Fish and Fisheries*, *15*(1), pp. 97–133.

Ryder, R. (1975) *Victims of Science*. London: Davis Poytner.

Sanders, C.R. (2006) "The sociology of human-animal interaction and relationships." *Ruminations*, *2*. https://networks.h-net.org/node/16560/pages/32228/sociology-human-animal-interaction-and-relationships-clinton-r-sanders last retrieved, 28 July 2021.

Searle, A., Turnbull, J. and Adams, W.M. (2022) "The digital peregrine: A technonatural history of A Cosmopolitan Raptor." *Transactions of the Institute of British Geographers*. https://doi.org/10.1111/tran.12566

Sebo, J. (2022) *Saving Animals, Saving Ourselves Why Animals Matter for Pandemics, Climate Change, and Other Catastrophes*. New York: Oxford University Press.

Seltenrich, N. (2018) "Down to earth: The emerging field of planetary health." *Environmental Health Perspectives*, *126*(7). https://doi.org/10.1289/EHP2374

Servais, V. (2018) "Anthropomorphism in human–animal interactions: A pragmatist view." *Frontiers in Psychology*, *9*, p. 2590. https://doi.org/10.3389/fpsyg.2018.02590

Shapiro, K. (2020) "Human-animal studies: Remembering the past, celebrating the present, troubling the future." *Society & Animals*, *28*(7), pp. 797–833.

Shipley, N.J. and Bixler, R.D. (2017) "Beautiful bugs, bothersome bugs, and FUN bugs: Examining human interactions with insects and other arthropods." *Anthrozoös*, *30*(3), pp. 357–372.

Shukin, N. (2009) *Animal Capital: Rendering Life in Biopolitical Times*. Minneapolis: University of Minnesota Press.

Singer, P. (1975) *Animal Liberation: A New Ethics for Our Treatment of Animals*. HarperCollins.

Sinha, A., Chowdhury, A., Anchan, N. S., and Barua, M. (2021) Affective ethnographies of animal lives. In *A Research Agenda for Animal Geographies*. Edward Elgar Publishing, pp. 129–146.

Tandon, R. and B. Hall (2014) "Majority-World Foundations of Community-Based Research." In Munck, R., McIlrath, L., Hall, B., and Tandon, R. (eds.), *Higher Education and Community-Based Research*. New York: Palgrave Macmillan, pp. 53–68. https://link.springer.com/chapter/10.1057%2F9781137385284_5

Taylor, H. (2016) "Marginalized voices: Zoömusicology through a participatory lens." In *Participatory Research in More-than-Human Worlds*. London: Routledge, pp. 52–67.

Taylor, N. and Twine, R., (2015) *Rise of Critical Animal Studies*. New York: Routledge.

Urbanik, J., (2012) *Placing Animals: An Introduction to the Geography of Human-Animal Relations*. Rowman & Littlefield.

von Essen, E., Turnbull, J., Searle, A., Jørgensen, F.A., Hofmeester, T.R. and van der Wal, R. (2021) "Wildlife in the digital anthropocene: Examining human-animal relations through surveillance technologies." *Environment and Planning E: Nature and Space*. https://doi.org/10.1177/25148486211061704

Waal, F. (1999) "Anthropomorphism and anthropodenial: Consistency in our thinking about humans and other animals." *Philosophical Topics*, *27*(1), pp. 255–280.

Wadiwel, D. (2015) *The War Against Animals*. Leiden: Brill.

Wadiwel, D. (2018) "Chicken harvesting machine: Animal Labor, Resistance, and the Time of Production" *South Atlantic Quarterly*, *117*(3), pp. 527–549.

Watson, A. and Huntington, O.H. (2008) "They're here—I can feel them: The epistemic spaces of Indigenous and Western Knowledges." *Social & Cultural Geography*, *9*(3), pp. 257–281.

Watt, S. and Wakefield, C. (2017) "Looking back. The use of visual methods in the social sciences." In Watt, S. and Wakefield, C. (eds.) *Teaching Visual Methods in the Social Sciences*. London: Routledge, pp. 8–23.

Weber, M. (1947) *The Theory of Social and Economic Organization*. New York: Free Press.

Wilkie, R. (2015) "Academic 'dirty work': Mapping scholarly labor in a tainted mixed-species field". *Society & Animals*, 23(3), pp. 211–230.

Wolch, J.R. and Emel, J. (eds.) (1998) *Animal Geographies: Place, Politics, and Identity in the Nature-Culture Borderlands*. London: Verso.

Wolfe, C. (Ed.) (2003) *Zoontologies: The Question of the Animal*. Minneapolis: University of Minnesota Press.

Xiao, X., Newman, C., Buesching, C.D. et al. (2021) "Animal sales from Wuhan wet markets immediately prior to the COVID-19 pandemic." *Scientific Reports*, *11*, 11898. https://doi.org/10.1038/s41598-021-91470-2

Part I
Theorising

2 Decentring Humans in Research Methods
Visibilising Other Animal Realities

Erin Jones and Nik Taylor

Introduction

In this chapter we outline how posthumanism has opened the door to consider other animals as subjects, and therefore (at least for some) legitimated their inclusion in the social sciences and humanities. However, we caution that simply positioning animals as entities to be studied risks objectifying them further, as they are frequently excluded from the research itself. While their exclusion is often a practical decision taken by researchers whose focus is how humans treat other animals, the fact remains that this decision epistemologically marginalises and thereby silences them, problematic for scholars who wish to advocate for other animals. In other words, the "master's tools will never dismantle the master's house" (Lorde, 2007 [1984]). We argue that a helpful corrective can be found by drawing on a feminist perspective that demands we critically interrogate how our methods might better line up with our theories. This leads us to advocate for a critical animal studies perspective that draws on expanded notions of intersectional feminism towards human-animal studies work. Using this as a framework, we discuss what it might look like to adopt methodological approaches that attempt to make humans reflect on their treatment of other animals. We do this through considering a case study where 13 human caregivers completed weekly journals designed to elicit the human's interpretation of their dogs' subjective experiences. While journaling has been used previously to consider human relations with other animals (e.g., Fox, 2006; Power, 2008), to our knowledge, it has not been used in studies that aim to decentre the human by encouraging an alternate animal-focused perspective. It is important to note that we did not intend the journals to give us, or our participants, a window into how dogs actually experience the world. Instead, they were devised to encourage empathy and intersubjectivity by prompting human participants to imagine their dogs' lives. The journals clearly showed that by being asked to take the perspective of their dogs, humans developed a stronger empathy for their animals, which often led to deeper reflection on how humans might help improve animal lives.

DOI: 10.4324/9781351018623-3

A Brief Note on Terminology

The field of human-animal studies is a broad one, and at its most basic denotes an interest from various disciplines across the humanities, arts and social sciences, in human relationships with other animals (DeMello, 2012; Taylor, 2013). Within this field, there are several differences that tend to turn on philosophical, ideological and political positions. Rhoda Wilkie (2015, 212) argues that these differences relate to: (1) the extent to which nonhuman animal scholars should be engaged with emancipatory-type research and (2) the emergence of the "animal as such–animal as constructed" axes. These broadly map onto the differences between critical animal studies (CAS), anthrozoology and human-animal studies (HAS) also referred to by some as mainstream animal studies (see Taylor & Twine, 2014). As we have argued elsewhere (see Taylor and Twine, 2014; Taylor & Sutton, 2018), we see anthrozoology as a field that rests on positivist ideals, claiming a scientific and/or neutral approach to the study of other animals. While we find claims of neutrality in any research spurious, those in anthrozoology often use this idea as a way of pointing out their research is not necessarily aimed at animal advocacy and/or that it eschews more radical ideas of animal rights and/or animal liberation. Critical animal studies scholars, on the other hand, embrace a more liberationist and advocacy position arguing that all CAS scholarship must be in service of activism for other animals. The main focus of such scholarships tends to be identifying and challenging the mechanisms that normalise and therefore allow the vast oppressions of other animals in our societies. These mechanisms are usually considered to be connected to, and supported by, capitalism in CAS scholarship. This focus on the mechanisms of oppression means there is much overlap with feminist animal studies (FAS), however, FAS is interested in interconnected oppressions between women, nature and animals, drawing on some aspects of ecofeminism (see Cudworth, 2016). While the term ecofeminism covers a broad school of thought (see, e.g., Adams & Gruen, 2014; Gaard, 2011) there are several key concepts that have been built upon by animal studies scholars. These include critiques of the culture/nature binary that devalues all associated with nature, namely women and other animals. The critiques also include the notion of interconnected oppressions, meaning that the mechanisms that support, normalise and encourage the oppression of one marginalised group are the same that support, normalise and encourage the oppression of other marginalised groups. In this vein, attempts to remove one form of oppression will not be successful without also dismantling the others.

The "Animal Turn"

Over the last couple of decades much has been written about the "animal turn", usually taken to mean the increasing attention given to other animals by scholars in the social sciences, humanities and arts (e.g., Mattes et al., 2021; Pedersen, 2019; Ritvo, 2007; Weil, 2010). While this increased attention from

scholars is largely welcomed, it is, as Steve Best (2009, p. 10) said "both lauda-
ble and lamentable". As Best argued, and one of us has argued elsewhere
(Taylor & Twine, 2014), moving animal studies from the margins to the centre
of academia comes at a price – an increase in a conservative, mainstream animal studies that fits (more) neatly within the academy. Or, as Best (2009, p. 10)
puts it, the "academic proponents" of mainstream animal studies "can only
advance it by currying for respect, credibility, and acceptance, which can only
come by domesticating the threatening nature of the critique of human suprem-
acism, Western dualism, and the human exploitation of nonhuman animals".

As a result of the general – although not wholesale – acceptance of mainstream animal studies in the academy, ideological differences have become
increasingly apparent in the field. This division is largely between those who
follow positivist injunctions to "objectively" study human-animal relations to
those who argue that no animal studies scholarship should come without
action and advocacy on behalf of other animals (Taylor & Twine, 2014). And
while this is an overly simplified characterisation and there are many who sit
at different points along the continuum between the two positions, it's fair to
say that the positivist works in animal studies that do little to challenge the
anti-animal status quo are far more numerous and well funded than the radical critiques (Taylor & Sutton, 2018). In part this is due to the popularity of
the animal turn in academia which has led to (much of) animal studies being

> co-opted and contained … muzzled and neutralized by a corporate-
> bureaucratic machine and its codes and logics. For once it takes shape
> within the sterile, normalising, hierarchical, and repressive environment
> of academia, animal studies, like any other knowledge or discourse, is
> tied to abstract, arcane, technical, and apolitical codes and discourses,
> and is reified as a marketable academic product and commodity as well.
>
> (Best, 2009, p. 10)

We see this in the increasing focus on the utility of animals to humans,
whether this be in terms of their support of our mental or physical health,
their ability to improve our "resilience" in the face of hardships, their impact
on human socio-emotional functioning, on child development and so on
(e.g., Morrison, 2007; Pachana et al., 2005; Putney, 2014). And while such
work is often couched in welfarist language about the need to "look after"
other animals properly, any consideration of the structural legitimation of
animal oppression is missing from this body of work. While this work has
gone some way in attempting to include other animals in academic considerations, it has done so by underlining the persistent and pervasive humanism
that sits at the heart of the western intellectual tradition (Taylor & Sutton,
2018) by having an innate anthropocentrism at heart.

That said, the field of human-animal studies has attempted to decentre the
human. This work has radical potential by re-imaging the very bases our
disciplines – and the western intellectual tradition – sit on (e.g., Peggs, 2012;
Ryan, 2011). This has forced scholars to revisit the gnarly problems of ethics,

ontology, language, power, epistemology, structure and agency, as well as questions of positionality and neutrality versus scholar-advocacy (e.g., Cudworth, 2016; Nibert, 2003; Peggs, 2013; Taylor and Sutton, 2018). This is partly because many of these attempts owe a debt to posthumanism, a philosophical turn that stresses the more-than-human configurations of social worlds (Ferrando, 2013). For some, this means it can be used in a critical way as the foundation of analyses of relations and systems of social domination where it is seen not as a rejection of humanism per se,

> but as a way of indicating our dissent from certain elements of humanism. While humanism is not a stable concept, we would want to reject humanist ideas that see the human species as in some way unique, exceptional, essential or distinct from the rest of nature.
>
> (Cudworth & Hobden, 2018, p. 8)

However, for many, it retains elements of a humanist legacy. While it stresses our interconnectedness and reliance on the more-than-human, this is seen as a different way to focus on "the human condition", in that it

> Actually enables us to describe the human and its characteristic modes of communication, interaction, meaning, social significations and affective investments with *greater* specificity once we have removed meaning from the ontologically closed domain of consciousness, reason reflection, and so on. It forces us to rethink our taken-for-granted modes of human experiences … by recontextualizing them in terms of the entire sensorium of other living beings.
>
> (Wolfe, 2010, p. xxv)

This is not the only problem in using posthumanism as a base to advocate for radical change for other animals. Possibly due to its intellectual grounding in Science and Technology Studies and Actor-Network Theory, posthumanism struggles to address structural bases and manifestations of power. It also tends towards the abstract with little clarity on precisely how it might be used to help liberate other animals. As Pedersen (2011, p. 75) notes,

> It should be clear that posthumanism's engagement with 'the animal question' does not in and by itself create more beneficial subject positions for animals in human society – it might, quite the contrary, obscure, dilute or displace responsibility for their situation, reinforcing rather than dismantling their exploitation.

Thus, even as posthumanism opens the door to, and at least partially legitimates, scholarly attention to human-animal relations, in order to develop a theoretical base for the emancipation of other animals we need to borrow from other theoretical approaches, namely an animal ecofeminism/feminist animal studies that draws on aspects of critical animal studies.

Concept box 2.1: Including non-human animals in research as subjects

As the study of our relationships with other animals has become more commonplace, there are numerous theories and ideas that we can borrow from. Many of these different theoretical approaches stress that animals are both actors and subjects in their own – and often our – lives. This idea of intersubjectivity is often at the heart of human-animal studies, critical animal studies, posthumanism and feminist animal studies. All of these fields stress a need to centre animals, a need to become less or non-anthropocentric in our thinking. However, scholars working in these areas have not necessarily considered how to do this in practice. In other words, while we may be making leaps and bounds in our thinking about the place of other species in our theoretical and conceptual apparatus, this is not matched by considerations of what it means methodologically. In part, this is because it's difficult to think about how we can include animals in our methods, and even harder to think about how we might do so ethically. Any attempt to make known the lives of others involves a re-presentation and difficult questions about power and knowledge. When one includes other species here – those who have historically been subjugated by dominant, normative methodological assumptions of humans – this becomes even harder: studying animals without (re)centring humans is a challenge. We need to develop specific methods to fully include other species in our research agendas and projects; methods that allow a sense of subject-hood that does not (solely) rely on humans speaking for other animals. This gives rise to some questions that are as follows: How can humans conduct research that avoids placing humans at the centre? How can animals' diverse interests, experiences and needs be understood and represented? How do we literally include animals in research projects without encroaching on their autonomy and free-will, given they can't provide informed consent? How can we develop methods that challenge us – as researchers and participants – to re-think oppressive human relations with other animals? Relatedly, what happens to our epistemologies and methodologies if we take the injunction to include animals seriously?

Including other animals in our research thus forces us to question often taken-for-granted assumptions about the centrality of humans (to the research process, as well as ontologically) and the mechanisms that marginalise nonhumans. In most disciplines this also requires consideration of the ideological and conceptual apparatuses that underpin them, as these apparatuses are normally always anthropocentric. In turn, this leads to questions about the role and functions of binary categories that pervade many disciplines (such as human versus animal). By and large, when it comes to human-animal distinctions, these binaries are operations of power that seek to relegate animals as simply "other than human". Finding ways to deconstruct, challenge and perhaps ultimately rectify this will have profound effects on scholarship.

Methodology in Feminist Animal Studies and Critical Animal Studies

While not always acknowledged (Adams, 2014 in Tuttle, p. 3) ecofeminist ideas about the role of dualistic paradigms in the maintenance of oppression and marginalisation pre-figure similar arguments from critical animal studies scholars (Fraser & Taylor, 2020). Similarly, both share a clear and specific focus on advocacy for animals. And while both schools of thought attend to the mechanisms that normalise human exceptionalism, ecofeminism leans towards considerations of "nature" while CAS focuses primarily on the role capitalism plays in the oppression of other animals. Perhaps due to these differences, paralleling the rise of critical animal studies, we have seen the development of a specific animal ecofeminism based on "the fundamental insight" of "the importance of speciesism as a form of oppression that is interconnected with and reinforcing of other oppressive structures" (Gaard, 2000, p. 206). In turn, this has led to the development of an explicit feminist animal studies (Cudworth, 2016) that overlaps with CAS but tends to have more acceptance and theoretical diversity bound by, as Cudworth (2016, p. 248) argues, "the appreciation of the precarious nature of animal lives, embodied materialism and a commitment to intersectional analysis".

Key to all of these positions, however, are (a) a clear position regarding the need for *action* and advocacy, (b) ideas of "total liberation" and species-inclusive intersectionality (Cho et al., 2013; Crenshaw, 1991, Kemmerer, 2011) and (c), an understanding that these first two concerns and closely linked to the fact that the methods we use are political, that is, they are implicated in how we see the world and who we see as legitimate subjects within it. As Sutton (2021, p. 377) explains when discussing radical emancipatory futures that take account of other animals,

> Research methods, which both shape and are shaped by the social world from which they arise, have the potential to contribute to this radical rethinking by visibilising realities that perpetuate or challenge dominant, humancentric, problematic ideas and highlighting new ways of being in the world with 'other' animals.

It is our contention here that choosing methods which either include other animals and/or allow us to advocate for them through the research process, and not just in their outputs, is key. It brings our research practices into line with our position as scholar-advocates and it enables us to practise emancipatory methods that contribute to research *for* other animals as opposed to simply research *of* other animals (see Peggs, 2013). In the remainder of this chapter, we discuss one of our attempts to do precisely this through the use of journaling by dog-caregivers.

The Journals: A Case Study

As part of a larger PhD research project (Jones 2019–2022) 13 human dog-caregivers were invited to take part in an experimental form of journaling. They were asked to keep a journal designed to encourage them to take the perspective of their dog for a six-week period. It is important to again emphasise that the journals were not intended to capture the dogs' experiences of the world, as that understanding is beyond us. They were, instead, aimed at prompting human participants to think differently about their dogs, to develop more empathy for them through encouraging intersubjectivity.

This concept of journaling was designed to – at least nominally — include the dogs' perspectives. While dogs were involved in the broader research project in other ways – e.g., through the use of photographs – the journals were specifically designed so that participants wrote entries in first person (dog) voice. Even though this offers a partial inclusion of other animals due to its reliance on human interpretation, it does visibilise the dogs' experiences. Furthermore, as we discuss below, it also leads to the human participants reflecting on their animals' lives in a different way, one which has the potential to improve them.

The journals were based on a technique of data collection commonly used in health care (Morrell-Scott, 2018), marketing (Siemieniako, 2017) and education research (Briscoe & Grabowsky, 2022). Journaling as a form of data collection can provide insight into the perceptions of participants in a variety of ways, with the "primary goal of journaling assignments to transform perspectives" (Briscoe & Grabowsky, 2022, p. 29). For example, our participants were asked to write from their dog's perspective as a way of including dogs and their perspective, but also of challenging them to think and write about their answers in a thoughtful, less human-centric way. The aim was to both provide data about how perspectives may influence interpretations of interactions and behaviour and "challenge dominant, humancentric, problematic ideas and highlighting new ways of being in the world with 'other' animals" (Sutton, 2021, p. 377).

Journal Content

Participants were asked to answer a series of six questions once per week for a six-week period. The six questions were repeated each week and were crafted to elicit the human's interpretation of their dogs' subjective experiences in both favourable and unfavourable interactions and activities based on relatively ubiquitous interactions (such as walks and training). For example, one question asked, "Did you get into trouble with your human(s) this week for doing something you shouldn't have been doing? Tell [us] about what happened and how this made you feel". Another asked, "Was there something you really wanted to do this week, but you weren't allowed to do? Tell [us] about what that was and how that made you feel". The questions were intended to be broad to allow participants to reflect on their own personal experiences rather than be directed to answer in a specific way. Although the expectation was that the responses would be highly metaphorical, the draw was to examine how these questions could broaden the ways people think about their dog's subjective experiences and build empathy towards their dogs' perspectives. In addition, the first and last weeks included opening/closing questions. Week one included five initial background questions asking participants about their dog and their history. On the sixth week, participants were requested to answer parting questions about their experience with journaling, exploring what was difficult about the process and what was helpful. For this chapter, we primarily analyse responses from the participants' final journaling assignment, although we do also include examples from weekly entries.

Participants were recruited through involvement in various dog-related groups on social media. Because of such a high interest in participation, the ads only remained posted for 2–3 hours. The journals were constructed using Qualtrics, a software platform designed to analyse user experience. Through Qualtrics, we sent participants an anonymous link with questions each week, along with mid- and end-of-week reminders. The questions were designed to elicit open-ended text responses, so participants could write as much or as little as they wanted. They were asked to create an alias for their dog and use that name as their identifying marker throughout the process. Participants were also invited, but not required, to submit weekly photos through an anonymous Dropbox link.

Animal Participants

The dog participants were as follows (names are aliases):

1 "Queeny" is a Staffordshire terrier and bulldog mix. She is two years old and was a "foster fail".
2 "Ella" is an unknown mixed breed. She is nine years old and was a rescue.
3 "Valentino" is a Border collie. He is two years old and was purchased from a breeder.
4 "Kody" is a Huntaway. He is three years old, was initially from a farm and purchased from Trade Me (an online marketplace serving New Zealand).
5 "Molly" is a Shih Tzu. She is five years old and was purchased from a breeder.
6 "Freddy" is a Chihuahua. He is 14 months old and was purchased from a breeder.

7 "Bubbles" is a Staffordshire terrier and greyhound mix. He is seven years old and was a rescue.

8 "Ellie" is a German shepherd. She is three years old and was purchased from a breeder.

9 "Lucy" is a Mastiff mix. She is ten months old and was a rescue.

10 "Patrick Stewart" is a Pitbull-type mix. He is a six-month-old rescue.

11 "Vashi" is a Boxer and a Pointer mix. He is 15 months old and was a rescue.

12 "Honey" is a Golden retriever. She is 18 months old and was purchased from a breeder.

13 "Meme" is an English pointer. She is six years old, was bred at home and stayed for life.

The content and length of entries differed across participants, but all focused on imagined dogs' perspectives (and were a delight to read). Examples included the following:

KODY: I love to dig holes in the backyard but Dad tells me off for this and I go in the naughty crate. I know I shouldn't do it and look really guilty every time.

LUCY: I really wanted to run free on the beach with the other dogs! Usually, I am allowed because I come back when they ask me to, but I didn't even get to touch the sand this weekend: I was distracted with a nice treat instead, and I guess I got over it pretty quickly, but it's really hard to watch all the other dogs go by and not get to have a little sniff at least!

QUEENY: The human child is ALWAYS cuddling me. I love it! She gives me a lot of pats and talks to me in a funny voice. I love getting praise it makes me feel good. I love getting treats too and the child is always wanting to give me treats but my human says no sometimes.

Re-thinking the Human–Canine Relationship through Journaling

Throughout history, animal narratives are present in stories and folklore, some exhibiting partially-constructed subjectivity that is both fantastical and intriguing. However, unlike fictional accounts of the inner lives of other animals, journals may offer something personal, deeply reflective, and a way to partially include other animals in the research that is about them. Additionally, because they challenge people to imagine the perspective of their dogs, they may also contribute to highlighting and challenging "the manifestations of power and oppression in these seemingly unproblematic everyday relations" (Sutton, 2021, p. 377).

In the remainder of this chapter, we focus on a discussion of how using journaling as a perspective-taking method evokes a reimagination of human-dog relationships for the human caregiver. We argue that such a method can contribute to an emancipatory praxis and shift in ethos both at the societal and individual level by visibilising nonhuman animal realities and challenging dominant human exceptionalist ideas.

Re-thinking the Human–Canine Relationship

The use of journals in the current research created a space for people to reflect on the subjective experiences of their dogs through a canine-centred approach. Borrowing from ecofeminist ideas about the role of dualistic paradigms in the maintenance of oppression, the journals aimed to decentre the dominant human exceptionalist influence and create space for other voices. Journaling was used as a way to encourage caregivers to re-think their relationships by fostering an awareness of their animals' lives through reflection of everyday interactions and by giving voice to their dogs. Participants were subsequently asked to describe their journaling experience and if their perspective changed over the course of the six weeks. It was clear from this data that the journaling method encouraged human participants to consider aspects of their dogs' lives they had not previously thought about. For example, Valentino's human wrote, "I already wrote a diary in Valentino's words - these questions have made me think more about how he might be feeling though, especially about learning. I have loved it!!!" Another participant wrote, "It's been quite fun to think in my dog's perspective. It's made me think a little more about how he feels and how things affect him".

Mindful reflection of the animals in our lives may emphasise how our own actions affect others in accordance with our dogs' subjective experiences, improving both their lives, singularly, and their lives living in relation to their human caregivers. For example, one participant wrote, "I became more aware of Molly's behaviour and how wonderful she is and how special she is. We tried new things and her biscuit eating is much better and more importantly, it's outside". Another wrote,

> [I]t has definitely made me more aware of my dog's week. It can be so easy to get wrapped up in what I am doing and not think about how that might make my dogs feel. Working full time and running a business means my two dogs sometimes miss out on a walk every day which isn't fair. The first question [in the journal] asking about their favourite walk has been a conscious reminder to walk my dogs 5–6 times a week and make sure they are really enjoying them.

It is this intersubjective awareness that can facilitate a (deeper) empathic connection which, in turn, might lead to the awareness of changes needed to improve animals' lives.

The intersubjective experience that journaling fosters builds empathy towards others, an empathy that is based on respectful acknowledgment of animals' capabilities. Aaltola (2013, p. 87) argues that through intersubjectivity "we presume that others have minds, that there are experiences and other mental contents with which we can identify". This intersubjective reflection is evidenced in the following journal entry, "I got more understanding of how easy it is to assume your dogs understand what you teach them. I think that I also wonder whether I know enough to understand how things I do affect my dogs". At a general level, this identification might serve to prompt weighty considerations regarding animals' rights, their relationship to us and our own sense of human exceptionalism. And at a specific level, identifying personal deficits in knowledge may benefit the way future interactions and teaching/ learning take place through a critical lens, in turn improving animals' daily lives. "Patrick Stewart's" entry reads, "These stories are always in my mind and when I look at my dog, I can 'hear' more stories to come. If anything, it makes me want to understand him better". This process may enhance the appreciation of canine companions that share our homes through forcing humans to reflect on the perspective of the animals they live with, as echoed by one participant who says, "I think I appreciate her even more".

Perspective Taking and Empathy

Across disciplinary boundaries and throughout time, empathy has been defined and constructed in many different ways. Young et al. (2018) define empathy as "a stimulated emotional state that relies on the ability to perceive, understand and care about the experiences or perspectives of another person or animal" (329) and includes the processes that allow us to take the perspective of another. Perspective taking can allow people to reflect on how their relationship may be impacted by taking an empathetic concern. One participant reflected that, "I think the general six weeks of notes made me feel differently about how my dog may be affected by what I do. I hope it has made me more empathetic".

Journaling also proved to be a way to influence the way participants mindfully chose to modify their week and activities based on their dogs' need and wants, centring their dogs purposefully and mindfully. For example, "If anything, knowing it was going to be journaled was maybe a bit more motivation to add variety and keep things interesting for her". Another example was provided by Valentino's human who says, "to be honest – he doesn't go for a walk very often due to our big property and I found myself planning what walk to take with him, in case I needed to write about it!" It is through this process that participants themselves were able to critically evaluate preconceived human-centric ideas.

Young et al. (2018) suggest that the most well-studied method for fostering empathy is through the imagination, which is partially what the journals intended. This occurs when we engage in perspective taking through reflection, storytelling, and role-playing, as was demonstrated using first-person canine narratives. Personal reflection inspires empathetic concern for those in our care and, in turn, people feel either good or bad about how the interactions are enacted. For example, one entry reads,

> [During the process] guilt arose because short cuts are unfair on a pack and that is my fault. It is my choice to have a pack so that has to be 24/7. I have put in better systems which are easier for all of us and we are more relaxed. I have enjoyed this, thank you, it was the shake-up I needed. I will give Meme more one on one time and stop the short cuts.

Anthropomorphism and Empathy

While, as we have argued above, certain methods have the potential to be emancipatory, we must still acknowledge that any animal-inclusive methods remain filtered through human understandings and therefore will inevitably be anthropomorphic in some way. One potential barrier to empathy is the use of anthropomorphism—a way of projecting that human characteristics to other animals, (or plants or object). Root-Bernstein et al. (2013) suggest that this inescapable human bias could be measured on a spectrum. At one end, people have a tendency to see other animals as unknowable beyond a human-centric moral concern, whereas at the opposing end, they are perceived to see experience the world in the same way humans do.

Anthropomorphism remains a contentious issue among those who research human-animal relations with it being clear that it offers benefits alongside potential harm. The construction of anthropomorphism can be extremely harmful because, as Horowitz and Bekoff (2007) say, "among lay people, anthropomorphism is not only prevalent, it is the nearly exclusive method for describing, explaining, and predicting animal behaviour". However, Bekoff (2000) also argues that "anthropomorphism allows other animals' behaviour and emotions to be accessible to us" (p. 867). Especially when considering emotions, Bekoff suggests that using anthropomorphism is mostly harmless, whereas "…closing the door on the possibility that many animals have rich emotional lives, …will lose great opportunities to learn about the lives of animals…" (p. 869). Though we acknowledge the journals were steeped in anthropomorphic metaphors, they also decentred human perspectives and made room for the idea that dogs have "rich emotional lives" in many ways. For example, "Molly" writes, "I came to mum today while she was sewing and put my front paws on the side of her leg to let her know I wanted a cuddle. I had to tap my front paws a few times until she picked me up. I liked sitting on her knee and seeing what she was doing. She stopped sewing and we went into the lounge and sat on the couch. I cuddled into her and looked up into her eyes and rubbed my head on her chest. We sat there for ages and I loved it".

Perspective Taking and Training

In this final section of the chapter, we turn briefly to discuss how journaling led to improved lives for some of the dogs involved in our study. We do this by focusing on how the journal entries led to a reconsideration of training techniques for the dogs. Training is about socialising dogs to navigate a human-centric environment. They learn human rules and skills that may help them thrive. Training can be an important part of the dog-human relationship, though it isn't always an ethical one. A multispecies perspective on the dog-human relationship, however, promotes embodied, empathetic beliefs of wellbeing and points in some useful directions. For example, according to Natalie Porter (2019) the dog-human relationship forces the consideration of wellbeing as it surfaces in bodies that are distinctly different from each other. A dog's behaviours and characteristics extend from markedly distinct foundations than those of their human caregivers. Thus, caregivers must learn to understand and appropriately evaluate dogs' body language to assess their motives, and only then intervene to shape those motives and behaviours (Yin, 2004). To achieve this type of interspecies communication requires that people teach themselves to anticipate and adapt to an individual dog's perceptual encounters of the environment (Haraway, 2003). It also requires that they move away from traditional dominance-based ideas about other animals, ideas that are often predicated upon unchallenged notions of human supremacy (Porter, 2019; Weaver, 2017; Włodarczyk, 2017). Journaling allows a reflection on training-based interactions that can lead to peoples' ability to consider their dog's point of view when teaching them. For example, one participant said, "I have learned a lot about how I train my dogs and how, what I do affects them".

What distinguishes the multispecies approach to the training relationship is its focus on how species differences engender limited similarities, disparities and ties (Latimer, 2013). This raises ethical concerns for the way we can learn from, interpret and choose to interact with and train dogs. As Porter (2019) says,

> So, while embodied interaction opens up new opportunities for mutual discovery and transformation as humans and animals perceive and respond to one another, it simultaneously raises critical questions about how we comport our own bodies in the presence of other species, and how to intervene ethically in and affect the bodies of others.
>
> (p. 104)

A method like journaling that allows people to reflect on such interactions may create a personalised learning space and foster understanding of how learning affects the ability for dogs to better integrate into society as co-occupants in shared spaces. Additionally, it may allow people to examine their own behaviour and how that is reflected in the type of educational interactions they choose. One participant writes, "Yes [journaling] did affect my perspective by making me think about how my dog may view her life. Understanding that helps a great deal with training and it has been a fun learning experience". It is worth a brief note here on our own position. We accept that human-pet relationships include a form of dominance, and we accept that training plays a role in this. However, we also accept that as currently configured, training – if done well with the animals' needs centred – can offer them a way to have more freedoms in societies that generally offer them very few liberties.

If we are to teach dogs the skills they require to live well in society with us, then teaching should be a process in which humans and dogs learn to become receptive partners through a two-way communicative relationship (Weaver, 2017, p. 10–11). The goal of teaching should be about providing optimism

(Balmford, 2017) and mental enrichment, something shown to be intimately linked with wellbeing. People can be made aware of the spaces lacking, whether it be gaps in skills and abilities, strengths and weaknesses or meeting their dog's individual needs and desires, and the data obtained from journaling can encourage this perception. For example, one participant writes,

> I don't think she gets enough mental stimulation – I thought she was but now I'm not so sure. So, we have been playing more 'games' and working on more operant games rather than classical and getting her brain going!

If the process of journaling is helpful in examining learning from a dog's perspective, this technique may not only open the doors to further investigation from scholars, but it could also be helpful for people to gain a better understanding about how their methods and interactions may affect their dog's ability to thrive in a human-centred society. As one participant writes,

> As a first-time dog owner, I like to think I am doing a lot of research into proper training techniques and being a responsible dog owner. Regularly thinking about things from her perspective can help keep you in check to make sure she is well looked after, happy, and well behaved.

The critical evaluation of techniques and perspectives is invaluable for challenging human exceptionalism.

Conclusion

In this chapter we have argued that posthumanism, if married with the more advocacy-based approaches of critical and feminist animal studies, has emancipatory potential for other animals. However, there is a clear need to ensure that the methods animal advocacy scholars use are relevant and play a role in visibilising other animals and challenging their treatment and position in society. The use of journals, that encourage a perspective taking, may begin to unravel anthropocentric beliefs about human exceptionalism by inspiring people to consider animals' subjective experiences. As we saw in earlier in the chapter, the use of journals in this way encouraged the human participants to think about life from the point of view of their animals, something we rarely do in our societies and communities. As such, it often led to changes in their behaviour aimed at making their animal lives better. Such an approach therefore has the potential to challenge oppressive entanglements that are often part of more traditional research methods and are very much a part of human-animal relations. It's feasible, then, that using this kind of method with others who live, work and/or interact with other animals might produce similar results. As a result, this kind of scholarship can be considered emancipatory as it moves beyond a traditional human contribution to a focus on empathy, inclusion and human behavioural and ideological change for the benefit of other animals.

References

Aaltola, E. (2013). Empathy, intersubjectivity, and animal philosophy. *Environmental Philosophy 10*(2), 75–96.

Adams, C. & Gruen, L. (2014). *Ecofeminism: Feminist Intersections with Other Animals and the Earth*. Bloomsbury.

Adams, C. (2014). Foreword: Connecting the dots. In W. Tuttle (Ed.), *Circles of Compassion: Essays Connecting Issues of Justice* (pp. 10–18). Danvers, MA: Vegan Publishers.

Balmford, A. (2017). On positive shifting baselines and the importance of optimism. *Oryx 51*(2), 191–192.

Bekoff, M. (2000). Animal emotions: Exploring passionate natures: Current interdisciplinary research provides compelling evidence that many animals experience such emotions as joy, fear, love, despair, and grief—We are not alone. *BioScience 50*(10), 861–870.

Best, S. (2009). The rise of critical animal studies: Putting theory into action and animal liberation into higher education. *Journal for Critical Animal Studies, 7*(1), 9–52.

Briscoe, G. S., & Grabowsky, A. (2022). The influence of journaling on nursing students: A systematic review. *Journal of Nursing Education 61*(1), 29–35.

Cho, S., Crenshaw, K., & McCall, L. (2013). Toward a field of intersectionality studies: Theory, applications, and praxis. *Signs 38*(4), 785–810.

Crenshaw, K. (1991). Mapping the margins: Intersectionality, identity politics, and violence against women of color. *Stanford Law Review 43*(6), 1241–1299.

Cudworth, E. (2016). A sociology for other animals: analysis, advocacy, intervention, *International Journal of Sociology and Social Policy 36*(3/4): 242–257.

Cudworth, E., & Hobden, S. (2018). *The Emancipatory Project of Posthumanism*. London and New York. Routledge.

DeMello, M. (2012). *Animals and Society: An Introduction to Human–Animal Studies*. New York: Columbia University Press.

Ferrando, F. (2013). Posthumanism, transhumanism, antihumanism, metahumanism, and new materialisms' differences and relations. *Existenz: An International Journal in Philosophy, Religion, Politics and the Arts 8*(2), 26–32.

Fox, R. (2006). Animal behaviours, post-human lives: everyday negotiations of the animal-human divide in pet-keeping, *Social & Cultural Geography 7*, 525–537.

Fraser, H., & Taylor, N. (2020). Critical (animal) social work: Insights from ecofeminist and critical animal studies in the context of neoliberalism. In C. Morley, P. Ablett, C. Noble & S. Cowden (Eds.), *The Routledge Handbook of Critical Pedagogies for Social Work*. London: Routledge.

Gaard, G. (2000). Review of Mary Zeiss Stange, Woman the Hunter. *Environmental Ethics 22*(2), 203–206.

Gaard, G. (2011). Ecofeminism revisited: Rejecting essentialism and re-placing species in a material feminist environmentalism. *Feminist Formations 23*(3), 26–53.

Haraway, D. J. (2003). *The Companion Species Manifesto: Dogs, People, and Significant Otherness*. Chicago: Prickly Paradigm Press.

Horowitz, A. C. & Bekoff, M. (2007). Naturalizing anthropomorphism: Behavioral prompts to our humanizing of animals. *Anthrozoös 20*(1), 23–35.

Kemmerer, L. (2011). *Sister species: Women, Animals and Social Justice*. Urbana, Chicago and Springfield: University of Illinois Press.

Latimer, J. (2013). Being alongside: Rethinking relations amongst different kinds. *Theory, Culture & Society 30*(7–8), 77–104.

Lorde, A. (2007 [1984]). The master's tools will never dismantle the master's house. In A. Lorde (Ed.), *Sister Outsider: Essays and Speeches* (pp. 110–114). Berkeley, CA: Crossing Press.

Mattes, S., Vincent, A., Whitley, C. (2021). Emerging with Oddkin: Interdisciplinarity in the Animal Turn. *Society & Animals 29*, 733–761.

Morrell-Scott, N. (2018). Using diaries to collect data in phenomenological research. *Nurse Researcher 25*(4), 26–29.

Morrison, M. (2007). Health benefits of animal-assisted interventions. *Complementary Health Practice Review 12*(1), 51–62.

Nibert, D. (2003). Humans and other animals: Sociology's moral and intellectual challenge, *International Journal of Sociology and Social Policy 23*(3), 4–25.

Pachana, N., Ford, J., Andrew, B., & Dobson, A. (2005). Relations between companion animals and self-reported health in older women: cause, effect or artifact? *International Journal of Behavioral Medicine 12*(2), 103–110.

Pedersen, H. (2011). Release the moths: Critical animal studies and the posthumanist impulse. *Culture, Theory and Critique, 52*(1), 65–81.

Pedersen, H. (2019). The contested space of animals in education: A response to the "animal turn" in education for sustainable development. *Educational Sciences 9*, 211.

Peggs, K. (2012). *Animals and Sociology*. Basingstoke: Palgrave.

Peggs, K. (2013). The "animal-advocacy agenda": Exploring sociology for non-human animals, *Sociological Review 61*(3), 591–606.

Porter, M. (2019). Training dogs to feel good: Embodying well-being in multispecies relations. *Medical Anthropology Quarterly 33*(1), 101–119.

Power, E. (2008). Furry families: making a human–dog family through home, *Social & Cultural Geography 9*(5), 535–555.

Putney, J. (2014). Older lesbian adults' psychological well-being: The significance of pets. *Journal of Gay & Lesbian Social Services 26*(1), 1–17.

Ritvo, H. (2007). On the animal turn. *Daedalus 136*(4), 118–122.

Root-Bernstein, M., Douglas, L., Smith, A. & Verissimo, D. (2013). Anthropomorphized Species as Tools for Conservation: Utility Beyond Prosocial, Intellectual and Suffering Species. *Biodiversity and Conservation 22*(8): 1577–1589.

Ryan, T. (2011). *Animals and Social Work: A Moral Introduction*. Basingstoke: Palgrave.

Siemieniako, D. (2017). The consumer diaries research method. In Kubacki, K., & Rundle-Thiele, S. (Eds.), *Formative Research in Social Marketing* (pp. 53–66). Singapore: Springer.

Sutton, Z. (2021). Researching towards a critically posthumanist future: on the political "doing" of critical research for companion animal liberation. *International Journal of Sociology and Social Policy 41*(3/4), 376–390.

Taylor, N. (2013). *Humans, Animals and Society: An Introduction to Human–animal Studies*. New York: Lantern Press.

Taylor, N., & Twine, R. (2014). *The Rise of Critical Animal Studies: From the Margins to the Centre*. London: Routledge.

Taylor, N., & Sutton, Z. (2018). For an emancipatory animal sociology. *Journal of Sociology 54*(4), 467–487.

Weaver, H. E. (2017). Feminisms, fuzzy sciences, and interspecies intersectionalities: The promises and perils of contemporary dog training. *Catalyst: Feminism, Theory, Technoscience 3*(1), 1–27.

Weil, K. (2010). A Report on the Animal Turn. *Differences: A Journal of Feminist Cultural Studies 21*(2), 1–23.

Wilkie, R. (2015). Academic "dirty work": Mapping scholarly labor in a tainted mixed-species field, *Society and Animals 23*(3): 211–230.

Włodarczyk, J. (2017). Be more dog: The human–canine relationship in contemporary dog training methodologies. *Performance Research 22*(2), 40–47.

Wolfe, C. (2010). *What is Posthumanism?* Minneapolis: University of Minnesota Press.

Yin, S. A. (2004). *How to Behave so Your Dog Behaves*. New Jersey: TFH Publications.

Young, A., Khalil, K. A., & Wharton, J. (2018). Empathy for animals: A review of the existing literature. *Curator: The Museum Journal 61*(2), 327–343.

3 Understanding human-animal relations from the perspective of work

Nicolas Lainé and Jocelyne Porcher

Introduction

Today, our relations with animals, and in particular with livestock, are widely criticised because of their impact on the environment, on human health, and on the animals themselves. The central driving force behind our relationship with animals, work, has been largely forgotten. However, work, this line of connection that brings together humans and animals, makes it possible to understand why we have lived with animals for thousands of years: *with* them, not *without* them.

This chapter offers an integrated approach to examining human-animal communities from the perspective of work. It aims to show that in the context of work, humans and animals not only partake in a shared world, but they create and transform this world together through collaboration. By extending concepts taken from the psychodynamics of work (Dejours et al., 2018) to nonhuman animals, this chapter presents an idea of subjectivity, where animals are beings that use skills and intelligence to collaborate with their human co-workers. Drawing on the notions of "conduct" and cooperation (see concept boxes in this chapter), we further argue that assumptions that animal work can only amount to alienation and exploitation are incomplete and limiting. We contend that the shared context of work is a space for relations between species, and that human-animal relations are not always a matter of humans dominating animals. We consider how techniques borrowed from participant observation or video recording are useful for an in-depth consideration of these relationships, and for shedding light on the animals' subjectivities. We draw on examples from our research on elephants in India (1) and farm animals in France (2) respectively. Finally, we conclude by highlighting the opportunities of focusing on animal and interspecies work for a better engagement of social science research with animal welfare and conservation.

Working with elephants in India: The Khamti and their elephants in northeast India

Upon arrival in India to pursue my PhD,[1] I was interested in exploring questions about the concrete relationships between humans and elephants.

DOI: 10.4324/9781351018623-4

I wanted to gain an anthropological understanding of the conditions and implications of the life that the local Khamti[2] share with elephants. Fundamentally, my aim was to shed light on what it means to share daily life with elephants: how does this work in practice? What ties have been established between humans and pachyderms? A particularity of interspecies communities is that both species have the same life expectancy and, consequently, living with an elephant is not the same thing as living in the company of a cat or a dog, as this shared life is inherently of a longer time horizon. The owners of the pachyderms and their mahouts very often spend several decades together. This leaves time for each of the protagonists to get to know the other, through sharing common experiences.

After some exploration of South and Upper Assam, I started conducting my fieldwork in the Lohit district in Arunachal Pradesh. I quickly realised that every man I met over the age of 20 was knowledgeable about elephants. What promising fieldwork for a young anthropologist! Indeed, in each of the households visited, at least one person had direct experience with these animals which they were able to tell me about. This also applied to my assistant Dipen, who, although only 19, knew a lot about elephants; he had grown up with them since boyhood, and he was able to mount and ride elephants. He often recounted his childhood memories of playing with the elephants captured by his father and elder brother, or of joining them in the forest for logging operations.

The Khamti have shared their daily lives with elephants, and worked with them[3], since they settled in northeast India. There are no official figures, but a brief census of informants made at the beginning of my research estimated the number of pachyderms in Lohit district to be nearly 500; this figure is particularly high in relation to the sub-continent as a whole. Yet despite the large number of pachyderms belonging to the Khamti, everyone, without exception, repeated to me that since 1996, most of their elephants have been sold. This was the year that an Indian Supreme Court decision banned logging and the sale of timber in the country. As this activity is the main occupation of elephants, as well as the main source of income for the Khamti, the latter were forced to sell their pachyderms. Today, while work in the timber industry has resumed, the feeling I have had since the beginning of my investigation, and which has subsequently been repeatedly confirmed, is that elephants are the one thing the Khamti lack the most today, and the cause of their absence, according to my informants, is the lack of work for them to do.

A study of the Khamti–elephant relationship's recent history clearly demonstrated the important role elephants have played in their society from the second half of the last century onwards. Since Indian independence in 1947, elephants have been essential for logging operations. At that time, the Khamti were mainly farmers, but with the increase in demand for elephants for logging, many of them began to specialise in the capture of pachyderms. Pachyderms, as we will see, occupy an essential place in timber operations, so they fast became the engine driving the Khamti economy, until the mid-1990s when the Indian Supreme Court banned logging and the sale of timber.[4]

However, the abrupt cessation of operations put both men and elephants out of work. Having no lucrative activity for their animals, and considering the economic value of each elephant, many owners sold them. The money raised by these sales allowed the wealthiest to invest in other activities (growing tea, bamboo, brick manufacture) and for those for whom the elephant was capital, the elephants provided a means of support for several years. Some did not wish to sell their elephants and continued to make them work illegally by ignoring the ban on logging for, despite the risks involved, they hoped for a rapid resumption of activities. The Indian Court had stipulated that their order and the prohibition it included would only be provisional. Each federal state, via its Forest Department, was to develop an appropriate forest use plan. Activities were officially resumed in the early 2000s, although in fact they had never really stopped.

This situation only exacerbated the socio-economic divisions between the Khamti living in villages and those living in urban centres. To cope with the economic precarity, those who owned elephants sold them. Although it is difficult to confirm, many Khamti told me that before the ban, each household had at least one or even two elephants.

In addition, this animal has always been used for domestic purposes (a means of transport for hunting, fishing, agricultural work, firewood collection, etc.). Today however, the use of pachyderms is largely limited to picking up and dragging logs. When the "Tai-Khampti Development Society" had a building constructed for its activities outside the city of Namsai, two Khamti lent their elephants before work began to prepare the ground for construction. This link between the animal–human relationship and interspecies work was therefore immediately central to my anthropological inquiries.

Understanding human-elephant relations through the theory of work

Unlike other animals, the idea that Asian elephants work is largely accepted. It is even one of the main arguments used by animal advocates. As they consider any form of work to be exploitation, animal rights advocates conduct campaigns aimed at removing certain pachyderms from their owners or the institutions that employ them. It should be noted that such a consideration of effective elephant work dates back to colonial times.

Nevertheless, approaching animal and interspecies work anthropologically was challenging at first. The disciplines that deal with the world of work within the humanities and social sciences often think of animals as simple production tools, assimilated to machines. This approach to animals in research is influenced by Marx's writings (1959), which clearly distinguish the work of men from that of animals. For Marx, work requires a mental operation, a consciousness that animals do not have, as they only operate instinctively. The work of animals has been excluded from anthropological research, as work is considered to be human alone, and therefore pertaining to the field of culture.

Since Marx, several researchers have tried to argue against the anthropocentric notion of work. Among them, sociologist Jocelyne Porcher hypothesised that animals are performing effective working activities (see the latter part of this chapter). Her research shows that in the relationship between farmers and farm animals, the death of the animal was ultimately only one aspect of their relationship. The rationality of the farmers' work with animals remains primary–not the economic aspect (making profit from the killing of animals). Thus, in work situations, the human-animal relationship takes precedence over the economic product (Porcher, 2002).

In her research, Porcher extended the psychodynamics of work theories, particularly those developed by Christophe Dejours (2009), to include animal work. Psychodynamics of work theories are concerned with the use of intelligence, the ways in which work is experienced, and its collective character. Research in this field has shown that in order to carry out the required task effectively, the work activity cannot be achieved by simply applying the procedures and regulations of the organisational framework within which it is carried out. In other words, there is no enforcement work. The psychodynamics of work distinguishes prescribed work from actual work. Prescribed work concerns all the rules and procedures that organise the work, and actual work concerns all the means invested by the worker to achieve the actual result. The work activity falls exactly into the gap between the prescribed and the actual. It involves "gestures, skills, a commitment of the body, the mobilization of intelligence, the ability to think, interpret and react to situations, it is the power to feel, think and invent" (Dejours, 2009, p. 20). According to Dejours, all these operations bring work to life, and, from work, we move on to the activity of "working". To do a "good job", as well as to overcome the constraints and suffering imposed by work, the worker must be capable of inventiveness, creativity and intelligence.

Moreover, the action of working is often, if not exclusively, performed within a collective framework. When an operation is performed, everyone is involved in a production activity. Productivity is not based on the usefulness of the individual, but on the fact that each individual works, and is involved in mutual productive activity; to work, you have to cooperate. Sociologists have shown (Cordonnier, 1997; Sabourin, 2012) that cooperation requires reciprocity between workers, that is, without the participation of one, the other will not cooperate. Christophe Dejours defines cooperation as "the way in which, collectively, workers reshape, reorganize, readjust and adjust coordination to make it efficient" (Dejours, 2013, p. 100). This definition clearly distinguishes cooperation from coordination. As it is only one dimension of cooperation, coordination "is limited to ensuring the articulation of singular activities" (Dejours, 2013, p. 101). In some respects, while coordination is, at the collective level, a requirement, effective collective work includes cooperation. This presupposes a willingness on the part of everyone to share and to deliberate on the task to be accomplished together, and this requires trust and visibility (Dejours, 2009). Further, when all the subjects have brought their intelligence together to form a common dynamic, they form a work collective, that is, "a team that has built its rules through such deontic activity[5] and 'space for deliberation', the place where [the rules] are discussed" (Dejours, 2013, p. 101). From a working individual, we move here to the collective: cooperation transforms work into "working with".

Thinking of animals as fellow workers means that they can no longer be considered passive beings at the disposal of humans, but rather, living subjects that are conscious, at least in part, of their actions. The psychodynamic approach places a greater focus on the invested means of achieving work than on its the final productivity; in other words, it seeks to highlight the way subjectivity and intelligence are engaged by animals to collaborate with human partners. In sum, in the psychodynamics of work, the question of working with animals means taking the individual and the relationship into account. It is not just a question of, on the one hand, studying the working conditions of people, and on the other, the working conditions of animals, it is a question of both. Thus, in my research, the elephant became, at work, a true companion to the mahout.

I first looked for signs of elephant investment in work. Work here concerns not so much its purpose, as the means invested to achieve it. It is therefore less a question of trying to show that the participation of pachyderms in activities is real work, than of looking at the way they do it. How do they use their bodies and minds and engage subjectively with men, to perform joint actions? What do these animals have to do with work? I also considered the nature of the ties forged with the men with whom they are engaged. What is the basis of their relationship with their mahout or owner? At work, do people and elephants cooperate and coordinate their actions? In other words, can we affirm that they form an interspecies work collective?

Methods and analysis

During fieldwork, the methods I employed were primarily ethnographic. First, semi-structured interviews and life-stories were collected from elephant owners and mahouts. These well-established tools helped to provide general information on the different uses elephants are put to and, more particularly, in knowing more about the impact of the 1996 timber ban.

Participant observation was then employed while accompanying elephants and mahouts in the forest. This method was supplemented with video recording, which rapidly became a useful tool for analysis [see DeAngelo's and Brown's chapters (Chapters 8 and 9) of this book]. Initially, these recordings were used as a means of observation, to supplement the notes and to remind us of the omissions when they were written up into ethnographic sketches. From a practical point of view, the use of videos recorded the time and date of the observations, and certain aspects of the context and the situation. The videos also made it possible to have backup for the interviews with the elephant capturers, which took the form of me asking them to explain what was happening between them and the animal during various activities.

These videos proved to be particularly useful for analysis. When I watched some videos on my return from the field, they proved to be the most effective way of observing the details, and the reactions of the elephants, which were sometimes difficult to notice at the time. These videos provided a means for emphasising the physical engagement between mahouts and elephants, and made it possible to note the frequency and nature of interactions. They revealed the sensory channels (visual, olfactory, tactile, aural) that were most employed when humans and animals communicate and coordinate their action. Using video as a medium for my ethnography allowed me to reveal a wealth of sensory information, such as the sounds or gestures of communication and the looks exchanged during the action. At another level of analysis, these videos made it possible to appreciate the investment of elephants in the accomplishment of their tasks (satisfaction, frustration), and the role of the emotional bonds woven with their mahout. This tool could therefore be used both for the presentation of observed facts and for their interpretation. For example, this is the approach undertaken by anthropologist and documentary filmmaker Natasha Fijn (2011) in her doctoral research into the links between herders and animals in Mongolia. In a recent article, she advocates the use of video in research involving humans and nonhumans (Fijn, 2012).

Results: working-with

In the forest, elephant/mahout work teams work together when a tree is felled and transformed into logs. Together, they have the task of transporting the trees from the place where they are cut to a previously defined gathering place. People and elephants then load them onto trucks that will transport the logs to the factory. For each of these activities, my observations and analyses show that each elephant knows what they are doing and what is expected of them (Lainé, 2016, 2018a, 2019). Moreover, among owners and/or mahouts interviewed there is no doubt that, at the very moment when men place ropes and accessories on elephants in the village, the elephants know where they are going, and what will be asked of them.

In order to successfully execute tasks, Khamti and elephants negotiate together to achieve a common goal. At work, all participants have the power to influence the outcome of the action being taken. While there is no doubt that the Khamti attribute intentions to elephants, the results of my research show that elephants also clearly recognise the people with whom they work (owner, mahout). Invariably, the shared knowledge of the pachyderms and the humans with whom they team up is fundamental to the achievement of tasks (Lainé, 2018a). All these elements make it possible to argue that mahouts/owners and elephants do not simply coordinate their actions in a "mechanical" way. The importance of autonomy and initiatives which are central to their relationship gives rise to real interspecies cooperation, in other words, to the formation of interspecies working groups (Dejours, 2013).

Video recordings allowed me to adopt an ethnographic approach that revealed the individual participation of elephants in work by showing how, in each of the activities in which they are involved, these animals make a personal contribution to the working activities. The results also show that working, or more specifically "working with", takes the form of a common beingness in shared living conditions (Lainé, 2016, 2018a).

Farm animals in France

Farm animals have mainly been studied in the context of "animal welfare", and with the aim of reconciling animal welfare with productivity. For more than thirty years, these studies have primarily involved experiments which entail placing groups of animals in different situations and comparing their behaviour and/or physiological parameters, based on hypotheses about the effects of one parameter on another. For example, to refer to an experiment in which I [Jocelyne] participated, the point under study was what effect different types of researcher behaviour had on a batch of 18 pigs. The animals were divided into three groups: the first were to be treated with kindness, the second with indifference, and the third was the control group. The results suggested that the pigs treated with kindness were more likely to interact than the others (Terlouw & Porcher, 2005). Since these experiments are financed by "animal welfare" programmes, it could be deduced from this conclusion that pig producers have an interest in showing kindness to their animals.

However, these experiments have several major conceptual and methodical flaws. First, they do not take into account animal subjectivity and intersubjective relationships between humans and animals (Porcher, 2002). When you interact with a pig, the pig asks themselves, and asks you: what do you want, and why? A pig is not a gas molecule or a bacterium. As soon as you are in the presence of an animal, communication is established. As Watzlawick said of humans, communication is inevitable, and this is no less true in our relations with animals (1964). In these experiments, there is a high probability that the pig will act according to what they think researchers want them to do. Pigs will not behave spontaneously, as the experiments are designed according to what the researchers want (and not according to what the animals may want). The other very important bias is that these experiments are carried out without the involvement of farmers – even though these are farm animals. Livestock farming is often referred to in the conclusions of articles reporting on the experiments, or even in the introductions, yet livestock farmers are not consulted and do not participate in the construction of the research. In other words, it is work itself that is the most absent from this farm animal research. Animals are considered outside the context of our relationships. However, it is work that gives meaning to our relationships and, without taking work into account in farm animal research, our interactions are reduced to meaningless behaviours.

As a consequence of recognising the failure to understand farm animals in terms of "animal welfare" in this type of research, I became interested in their relationship at work, and began by asking the question: do animals collaborate at work? (Porcher & Schmitt, 2012). Then, having answered this question positively, I followed up with the question: do animals work? (Porcher, 2017b). Understanding the relationship between animals at work means taking an interest in their capacity for subjectivity, intelligence, initiative and resistance. How can animals be studied without using experimentation, and no longer from the point of view of biology and the natural sciences, but rather from the perspective of the social sciences? With which methods? With which tools?

By drawing on the psychodynamics of work (presented in the first part of this chapter), I researched workplace suffering in industrial systems to better understand human relationships at work (Porcher, 2006). By observing animals at work, it seemed to me that this theoretical framework could also help us understand an animal's behaviour and the issues at work for them, and for us.

This is why in 2007, in collaboration with Tiphaine Schmitt, I created the first field study of relationships between working cows and farmers (Porcher & Schmitt, 2012). Tiphaine conducted participant observation for three months on a dairy farm. She interviewed farmers, and we built a protocol for live and video observation of the animals. In other words, we used methods similar to those we would have used if we had observed children in a family or at school. Animals, like young children, do not talk, but their actions speak for them. The results showed that cows collaborate at work, or rather, that they try to be as uncollaborative as possible. The farm on which we did our research was not particularly pleasant for the cows. It was a zero-pasture farm, with high productive pressure and a rather violent producer. In these difficult conditions, cows were nevertheless obliged to work, just like a factory worker, even though the work has no economic interest for the cows. We have shown how cows value work and why this type of intensive system is counterproductive, as it deprives itself of the animals' intelligence and ability to act.

I have recently undertaken research which is also based in the psychodynamics of work, with another colleague, Sophie Barreau, at a training centre for breaking in young race horses (Porcher & Nicod-Barreau, 2019). Our hypothesis is that this breaking-in period corresponds to a professional training that prepares the animal for its future profession. We think of the horse as an actor in their training and we study their relationships at work. As the training follows a very standardised 20 stages, we have been able analyse the evolution of the horse's behaviour step by step, as well as the behaviour along the way. Video has been a central tool for these observations, and Sophie has filmed a hundred horses during these twenty stages (4000 videos –mean 10 mn). Furthermore, we studied certain stages using the Observer XT software[6] with the objective of highlighting occurrences of behaviour and differences between variables (horse, trainer, age of the horse, sex of the horse). The first results of this ongoing research show the importance of affectivity in the relationship between trainers and horses, and the importance of voice in the training work. These results confirm those obtained elsewhere (Lainé, 2016; Porcher & Estebanez, 2019). In working relationships with animals, voice is the bearer of intentions and affects, and is an important medium of communication at work. Within the context of training young horses, the voice can emotionally envelop, reassure, encourage and convince. The trainer's goal for their voice is to convey confidence that the animal can trust the human being.

The question of animal work

In both of the case studies presented above, the question of animal work offers a crucial alternative to counter the criticisms of an anthropocentric view of animal domestication, and it presents a new perspective on interspecies relationships (Porcher & Nicod-Barreau, 2018). Investigating human-animal relations could not only illuminate sustainable farming systems but could also offer new insight into species conservation (Lainé, 2018a) and biocultural diversity (Lainé, 2018b).

The question of animal work is often dismissed, for do animals indeed work? To dismantle this attitude, it was first necessary to go past the idea of work as domination of man over animals. By choosing to observe conduct rather than behaviours, it is possible to verify whether the animals under observation have understood the task required of them, and to identify elements that indicate their ways of engaging and disengaging.

Concept box 3.1: Conduct rather than behaviour

As demonstrated by the psychodynamic of work, "working" is apparent from the worker's subjectivity. To do a "good job', but also to overcome the constraints and sufferings imposed by work, the worker must be capable of inventiveness, creativity and intelligence.

To achieve this, the analysis of the different tasks performed by animals must be considered through the notion of conduct, understood as "way of acting which presume intelligence and action" (Porcher & Schmitt 2012, p. 241). The notion of behaviour, although recently debated in a critical and transdisciplinary framework (Burgat, 2010), refers to a certain form of ambiguity. Emerging from the discourse of behaviourists, and still widely used by animal sciences today, behaviour does not presuppose the intelligence of the individual. On the contrary, in its most Pavlovian sense, its use would risk interpreting each of the actions of animals in terms of conditioning, which would give humans the primacy of their actions over animals. Interpreting the animal's actions in terms of conduct makes the animal an active agent, that is to say the subject is not acting influenced by its genes or by conditioning, but the animal is an actor in what he or she does (Porcher, 2017a).

Thus, employing conduct rather than behaviour allows us, on the one hand, to discuss whether the observed animals have understood the task required of them (or whether they simply act as beasts of burden, responding to commands dictated by humans); on the other hand, to identify elements that indicate the animals' commitment (animal intelligence, consent, renunciation, affectivity…).

The psychodynamics of work draws into focus several aspects of what it means to be collectively engaged in a common activity. At work, do individuals form collectives and cooperate (i.e. by engaging their own subjectivity), or do they simply coordinate their actions? This research poses a challenge, because coordination implies a mechanistic idea of the animal, while active cooperation requires the use of intelligence to accomplish a subjective task. The methodological approach is to rely on methods which can collect observational data by paying close attention to the operations performed by humans and animals.[7]

Concept box 3.2: Coordination and cooperation

Since one is never working alone, coordination is a necessity. A foreman or a project manager coordinates the work; i.e., they distribute the tasks, directs who works with whom, with what means and with what objectives. Coordination is part of the process of work. It connects working people, and accepting this coordination is part of the subordination relationship linked, for example, to paid work. On the contrary, cooperation cannot be decreed. It depends on the will and desire of people and is deeply linked to the intersubjectivity of relationships. Cooperation is about engaging beyond the prescribed rules with someone to do something. This may be what managers are looking for, but is not necessarily achieved. In the case of animal labour, we can also distinguish between coordination and cooperation. In the example of the dairy farm discussed in this chapter, the cows responded to coordination orders, such as going to the milking robot when requested by the producer. But they were not cooperating: if the producer turned his head, they could change direction and not go to the robot. They did not want to deal with the producer. He was assured of their relative obedience, but not of their cooperation.

Conclusion

In this chapter, we have offered an account of how participant observation and the use of video offer useful tools for exploring human-animal relations in the context of work. Human work has historically been the realm of the social sciences, while animal work is a promising research avenue in human-animal studies. Animals are currently being appropriated by supporters of industrialism,[8] or of welfarism, or of abolitionism, and the topic of animal work is not a simple matter, considering the diversity of these agendas built on economic and social history of our relations with animals. The work of animals is still a relatively new research arena. This research allows us to test our capacities as researchers to use, and even invent tools that enable us to ask animals the right questions – questions that facilitate our understanding them, as opposed to questions which confine them. As Dejours writes, work contains the promise of emancipation. It is up to us to make sure that such promise is finally upheld.

Notes

1 The data presented here was collected between 2008 and 2010 during my Ph.D. fieldwork in Arunachal Pradesh (Lainé, 2014, see also Lainé, 2020).
2 The Khamti are an ethnically Tai community (tai-Kadai Branch) who migrated towards northeast India at the end of the 19th century.
3 The proximity between the Khamti and elephants was possible because of the presence of large numbers of pachyderms, which are an integral part of the ecosystem of the region. North-east India is an excellent natural habitat for elephants. Located on the foothills of the Himalayas and intersected by the Brahmaputra River, which winds through the valley for nearly 600 km from north to south, as well as its many adjacent rivers, the region contains a series of hills which serve as refuges for packs of elephants that retreat there during the annual monsoon season (Lainé, 2012). The ties between humans and nonhumans have been documented since ancient times and, according to the latest census, the region is one of the most important population centres for the species worldwide. Today, despite the consequences of development, particularly in the southern valley, the Northeast is home to the largest wild elephant population in Asia (*Elephas maximus*). Nearly 10,000 individual animals were counted in the early 2000s (Bist et al., 2002).
4 This decision led to economic impoverishment and social disintegration within Khamti society. During the course of the study, I observed significant opium consumption. It was present every day during our talks, as well as in the forest camps (see Lainé, 2012).
5 According to Christophe Dejours, a deontic activity, or "deontic of doing", refers to the links forged between subjects in order to work together (Dejours, 2009).
6 The Observer XT software allows the collection, analysis and presentation of observational data.
7 It is useful here to remind the reader that the notion of collective of work reflects "a team that has built its rules through such deontic activity and 'deliberation space', the place where they are discussed" (Dejours, 2013, p. 101). Thus, the notion of cooperation is more crucial and more pertinent than coordination.
8 Whether it is the heavy industry of animal production inherited from the 19th century or the cellular agriculture that is currently being set up, as in vitro meat for example (Porcher, 2019).

References

Bist, S.S., Cheeran, J.V., Choudhury, S., Barua, P., Misra, M.K., (2002), 'The domesticated Asian elephant in India: Country report', *in* Baker, L., Kashio, M. (eds.), *Giants on our hands: Proceedings of the international workshop on the domesticated Asian elephant*, FAO Regional Office for Asia and the Pacific, Bangkok, pp. 129–143.

Burgat, Florence. (2010), *Think animal behaviour: contribution to a critique of reductionism. Comment penser le comportement animal?* Editions Quae, Paris, France.

Cordonnier, L. (1997), *Coopération et réciprocité*, Presses universitaires de France, Paris.

Dejours, C. (2009), *Travail vivant—Tome 1: Sexualité et travail*, Payot, Paris.

Dejours, C. (2013), *La panne*, Bayard, Paris.

Dejours, C., Deranty, J.P., Renault, E., Smith, N. (2018), *The return of work in critical theory*, Columbia University Press, New York.

Fijn, N. (2011), *Living with herds. Human-animal coexistence in Mongolia*, Cambridge University Press, Cambridge.

Fijn, N. (2012), 'A multi-species etho-ethnographic approach to filmmaking', *Humanities Research*, XVIII(1): 71–88. DOI: 10.22459/HR.XVIII.01.2012.05

Lainé, N. (2012), 'Effects of the 1996 timber ban in Northeast India: The case of the Khamtis of Lohit district, Arunachal Pradesh', *in* Lainé, N., Subba, T.B. (dir.), *Nature, environment and society: Conservation, governance and transformation in India*, Orient Blackswan, New Delhi, pp. 73–93.

Lainé, N. (2014), 'Vivre et travailler avec les éléphants: une option durable pour la survie de l'espèce. Enquête sur les relations entre les Khamti et les éléphants dans le Nord-Est indien', PhD Dissertation, Université Paris-Ouest Nanterre, Nanterre.

Lainé, N. (2016), 'Pratiques vocales et dressage animal. Les mélodies huchées des Khamtis à leurs éléphants', *in* Bénard, N., Poulet, C. (dir.), *Chant pensé, chant vécu, temps chanté: formes, usages et représentations des pratiques vocales*, Éditions Delatour, Paris, pp. 187–205.

Lainé, N. (2018a), 'Coopérer avec les éléphants dans le Nord-Est indien', *Sociologie du travail [En ligne]*, 60(2) (Avril-Juin 2018), online on 24th May 2018, consulted on 4th September 2018. URL: http://journals.openedition.org/sdt/1953; DOI: 10.4000/sdt.1953

Lainé, N. (2018b), 'Asian elephants conservation: Too elephantocentric? Towards a biocultural approach of conservation', *Asian Bioethics Review*, 10(4): 279–293.

Lainé, N. (2019), 'For a new conservation paradigm: Animal labor. Examples of human-elephant working communities', *in* Porcher, J., Estebanez, J. (eds.), *Animal labor. A new perspective on human-animal relations*, Transcript – Verlag, Paris, pp. 81–101.

Lainé, N. (2020), *Living and working with giants: a multispecies ethnography of the Khamti and elephants in Northeast India*, Muséum national d'Histoire naturelle, Paris.

Marx, K. (1959), *Le Capital*, tome 1, Éditions sociales, Paris.

Porcher, J. (2002), *Éleveurs et animaux: réinventer le lien*, Presses universitaires de France, Paris. DOI: 10.3917/puf.porch.2002.01

Porcher, J. (2006), 'Well-being and suffering in livestock farming: Living conditions at work for people and animals', *Sociologie du travail*, 48(1): 56–70.

Porcher, J. (2017a), *The ethics of animal labor. A collaborative utopia*, Palgrave Macmillan, nom de ville?, Cham

Porcher, J. (2017b), 'Animal work,' *in* Kalof, L. (ed.), *The oxford handbook of animal studies*, Oxford University Press, Oxford, pp. 302–318.

Porcher, J. (2019), *Cause animale, cause du capital*, Editions Le Bord de l'Eau, Lormont.

Porcher, J, Estebanez, J. (eds.), (2019), *Animal labor. A new perspective on human-animal relations*, Trancript Verlag, Bielefeld. Germany.

Porcher, J., Nicod-Barreau, S. (2018), 'Domestication and animal labour', *in* Stepanoff, C., Vigne, J.D. (eds.), *Hybrid communities. Biosocial approaches to domestication and other trans-species relationships*, Routledge, London.

Porcher, J., Nicod-Barreau, S. (2019), 'Le débourrage des jeunes chevaux. Un terrain inattendu pour la psychodynamique du travail?' *Travailler*, 41: 153–169.

Porcher, J., Schmitt, T. (2012), Dairy cows: Workers in the shadows? *Society & Animals*, 20: 39–60. DOI: 10.1163/156853012X614350

Sabourin, E. (2012), *Organisations et societes paysannes. une lecture par la réciprocité. Une lecture par la réciprocité*, Editions Quae, Paris.

Terlouw, E.M.C., Porcher, J. (2005), 'Repeated handling of pigs during rearing. I. Refusal of contact by the handler and reactivity to familiar and unfamiliar *humans'*, *Journal of Animal Science*, 83(7): 1653–1663.

Watzlawick, P. (1964), *An anthology of human communication*, Science & Behavior Books, Palo Alto, CA.

4 Re-thinking Animal and Human Personhood

Towards co-created narratives of affective, embodied, emplaced becomings of human and nonhuman life

Owain Jones

Their [animals'] whole being is in their living flesh.

(Coetzee, 2003, p. 110)

I see animals, inside the people, I see people, inside the animals.

(Al O'Kane, 2015)

Introduction: Shared Life

The proposal in this chapter is that nonhuman animal becoming and person-hood are articulated in particular, affective, spatialised, relational, embodied and precise practices of living individual creatures. By studying the embodied lives of both animals *and* humans, one can come to realise the following: that animals do actually have personhood; that there are many threads of shared becoming between animals and humans; and that this is often expressed in affective registers. As a result, the divisions that exist between humans and animals in modern systems of thought, including human exceptionalism and concerns about anthropomorphism, are not justified.

To say animals have personhood immediately opens up controversial ideas in science and philosophy in recent decades and also reach back into older debates. Modern thought systems and the economies, cultures, politics and ethics that underpin these systems have drawn sharp distinctions between humans, animals and nature more generally. Humans are seen as special, as different in *kind* and value, to the rest of nature. In part this comes from the undoubted power of the human brain to know, remember and imagine, to have a narrative of self and of the world from an individual perspective.

In part, the separation of modern human from nature also came from other sources, particularly, through some religious ideologies which say that humans were created "in the image of God" in ways in which the rest of nature was not. For example, some environmental scholars point to the idea of *dominium terrae*, the idea that God grants dominion over the earth to "man", as a key source of the environmental crisis (see, e.g., White, 1978). This idea occurs in a number of the monotheistic religions. In Christianity it occurs very early on in the "story of creation" featuring in *Genesis* chapter 1: verse 28.

DOI: 10.4324/9781351018623-5

Giorgio Agamben diagnosed the "history of both science and philosophy as part of what he calls the 'anthropological machine', through which the human is created with and against the animal" (Oliver, 2007, p. 1). At the heart of this machine is the perceived and claimed difference between humans, and animals/nature, which is best captured by the term "human exceptionalism". Humans are seen as separate from, and superior to, nature. Humans are seen as valued individual persons with self-identity. This then leads, in modern philosophies and politics, to ideas of inalienable human rights, including the right to live in freedom as an individual (as in the United Nations Universal Declaration of Human Rights that was decreed in 1948). Nature, and within that, animals, in contrast, are seen as things dominated by instinct, without self-awareness or individuality and personhood, and thus not meriting the same rights as humans.

Scholars from a range of disciplines now argue three things: firstly, the sharp distinction between humans and animals, in particular, is not a true reflection of the lived realities of either; secondly, human exceptionalism is one of the fundamental causes of the terrifying and existential global environmental crisis we now face; thirdly, under this system of thought, animals are systematically exploited and abused and their rights denied.

While some human cultures have lived within nature, most obviously some indigenous first nation cultures, many others became separated from nature in spiritual, ethical and political terms through the development of the already mentioned theologies and the technologies and economies of agriculture, industrialism and capitalism. This move away from nature seems to have come to a climax in the societies which have colonised and exploited the earth in the last six centuries or so. The industrialised world in particular has been organised to serve human needs and desires at the expense of nature, using nature as a resource for a separated human society, nearly always putting human priorities over natural ones in decisions of development. For example, this is expressed in the continuing cultivation of wild lands for development, agriculture and other forms of production; and the dumping of human-generated waste into the atmosphere and oceans with no regard for the consequences.

Another important point relevant to my subsequent arguments is that through the Enlightenment project of modern knowledge, which emphasised rational and objective science, humans were understood as separate not only from nature, but from their own (natural) bodies. This is known as the Cartesian Mind–Body Dualism.

Concept Box 4.1: Cartesian Mind–Body Dualism

The mind–body dualism is one of the fundamental elements of the philosophy of Rene Descartes, whose ideas were key in the development of the Enlightenment, rationalism and science. He considered the mind to be of a different order of existence and substance to that of the body. He also considered the mind's mental conscious capacities to be the locus of human nature, value and agency. This focus on the human (individual) mind as the core of human identity and value is now widely seen as one of the fundamental drivers of many environmental problems, for example, the poor treatment of animals and the exploitation of land.

The uniqueness of humans, as a species and as individuals, our personhood, the very core of our being, was seen to be in the thinking mind. Descartes understood the body as merely a mechanical object to be operated by the mind. The body was, in fact, considered animal. Thus, following the valuing of the mind over the body as the locus of human identity, value, personhood and rights, it was easy to disregard animals as bodies without (rational, human-like) minds.

No one denies that animals have bodies. In fact the reverse, as we use and value (economically) animal bodies and body parts for all manner of commodities - on industrial scales. But within Enlightenment and Cartesian thinking, it was easy to deny that animals had minds and thus personhood, moral value and rights. This left them open to the justification of many forms of exploitation and disregard that are so evident today.

In recent decades, animal rights theories, animal cognitive and behaviour science studies and certain branches of philosophy have challenged the idea that animals do not have senses of self and thus personhood (Rowlands, 2019). Work in animal geographies have drawn upon, supported and extended this direction, paying particular attention to the spatiality of animal life. Here I seek to contribute to this by thinking about how animals, and individual animals, actually live, and how humans actually live, as *embodied beings in places and times*.

I suggest that there are many shared traits of becoming between individual humans and individual animals ("becoming" meaning being that is always unfolding into the future through time and space in novel negotiations, albeit carrying legacies and trajectories from the past). Immediately following the notion of individual becoming comes the notion of "becoming-with" (Haraway, 2008). This means that all organisms live shared, relational lives, with others of their kind, with other species and their environment. This is, in fact, what ecology studies: connected life. Connections between living human and nonhuman beings have been effectively considered through notions such as intersubjectivity by Barbara Smuts (2001) and Vinciane Despret (2004).[1]

It could be argued that the idea of (human) individual becoming is trapped in the dangerous delusion of the enlightenment/modern notion of the free individual. All beings live in a mesh of ecological connections in terms of, for example, nutrients, air, water and waste. We are always with other beings, other forces, other elements: that is the shared basis of *all life*. It is very hard for modern humans to abandon the idea of individuality and freedom so somehow, it persists while we seek new ways of imagining human life, and human life with nature. For example, Andreas Weber has recently coined the idea of aliveness as a shared form of becoming between humans and nature, in his advocacy for a new post-enlightenment poetics for a liveable Anthropocene:

> I propose we understand the identity of humans and nature through a commons of creative transformation that underlies all reality and finds a particularly forceful expression in life. [T]here can be no dualism, because the fundamental dimension of existence is already shared: it is *aliveness*, the desire to connect through touch and body in order to create fertile communities of mutual flourishing, the members of which experience their identities as selves.
>
> (Weber, 2019, p. 3; emphasis in original)

No being or organism is free in terms of being independent from all the systems and processes we are ecologically enmeshed in. Becoming is always becoming-with through that ecological embeddedness which is utterly shared with others, both human and nonhuman. I seek to show how, through some relatively simple "thought" experiments based upon everyday observations, we can begin to explore these kinds of ideas. The aim is not to eradicate all difference between humans; between humans and nonhumans, and between different nonhumans, but rather, *to put difference on a different footing*.

There are, of course, many registers of difference between both human and nonhuman becoming, in body, brain, habit, habitat, environment, ecological matrix, but there are also many threads of shared, affective life in all these becomings-with, which bind them as the family of life on earth. The suggestion I am making is that the performativities of bodies in everyday life (human and animal) can be *mobilised as ways of knowing of animal becoming and personhood*. This is rather than assuming that individual personhood, if it exists, is formulated in inner thoughts and emotions which are very hard to read by an observer.

Examples from various sources are shared later, but here is a simple example, to start with – stretching. Think of stretching when you wake in the morning, or pause from reading for a moment now, and really stretch your arms out, push your shoulders back, and feel the feeling. What is happening in your body/mind? All humans stretch. Most animals stretch. Stretching is a useful, complex mind–body function, but it is, in part, a moment of transition from one body state to another – or a pause in body state for relief. When you stretch you are affectively in a stretching-becoming moment, with a view to what comes next. When a cat stretches (as they do so beautifully), he or she is in a shared state with your stretching.

Animals also yawn, stroll, sprint, feel cold, hunger, thirst, anger, anxiety, boredom and arousal. These are all shared states of becoming which shape lived becoming as it unfolds. This profound idea reaches all the way back to Charles Darwin's third great, but much less referred to work, *The Expression of Emotions in Man and Animals* (1872) in which Darwin explored the continuities between human and animal life (see Figure 4.1). The great philosopher of feeling, Spinoza felt animals had some shared affective capacities with humans. This early work on animal life did not however strongly

Figure 4.1 Shared affective states between humans and animals.

Illustrations from Charles Darwin's *The Expression of Emotions in Man and Animals* (1872).

challenge the idea of human exceptionalism or lead to empathy for animals, or interests in animal rights. Hasana Sharp (2011) points out that in "denying the affective *community* we share with nonhuman animals, Spinoza overlooks the joyful and enabling features of our proximity to them" (2011, p. 66).

The following section of the chapter explores in more depth how animals have personhood through their embodied and spatial becomings, that there is much common ground between animals and humans and that we are unified through the process of being alive.

The Placedness and Embodiedness of Animal and Human Life

My suggestion is that animals' lives are differently materialised, socialised, spatialised and *performed* in individual life narratives acted out in affective relation (with others and environment) in place, as well as sketched out, or enframed, in their evolutionary, genetic, behavioural repertoires.

In making this suggestion I draw inspiration from a number of sources. These include growing up on a farm in the UK which teemed with cows, sheep, pigs, hens, geese, dogs, cats, rats, flies, spiders and many types of wild birds. And, as an adult, living for about four decades with three generations of beautiful cats sharing our house. All these fabulous feline characters have been very different in personality, habit and interactions with us, each other and their environment. Of note, and delight, have been the bewilderingly subtle and varying geographies of where to sleep, both inside and outside, that each cat has lived out. I also draw upon animal geographies and a range of writings (novelistic, nature writing, social/natural scientific, philosophical) which, in differing ways, try to get close to animal becoming, and share the idea that individual animals have differently articulated forms of becoming. These often tend to be texts pushing at the imaginative borders of animals' worlds and human-animal worlds.

All living creatures are affective, *spatialised*, becomings-with. I suggest that geography, as the spatial discipline, has an intellectual, methodological, moral and political duty to continue to engage with, and pay the closest witness to, animal lives. Animals' lives are geography's trade – dealing with/in space. The same goes for human becoming too, and recent human geography has increasingly focused on how human daily life is enacted, emplaced and embodied affectively (see, e.g., Silvey, 2017).

The possibility of considering animal becoming in place, and in shared lives, presents both great challenges and opportunities to the coming generations of scholars in human-animal studies, not least because modern human relationships with animals and wider nature are so important and, in many ways, so problematic. Another thing to bear in mind is that the famous philosopher of place, Edward Casey, has suggested that place itself has been a disregarded or even oppressed idea and value in modern Western thought (Casey, 1997). Thus, thinking how humans and animals live embodied lives in places means challenging many powerful, joined up, oppressive traditions of thought.

Once animals are recognised as having personhood in place, and are individuals, many sharp questions arise. Philosophers such as Peter Singer have long discussed issues of individual animal rights (see, e.g., Singer, 1975). Recent debates within (human) geography have added important strands of thought to these ongoing questions of human becoming, human exceptionalism, animal becoming, animal rights, sustainable – or otherwise, futures for collective life on earth.

In a key work of animal geography, Philo and Wilbert (2000, p. 5) asked, "can a 'real' geography of animals be developed" – that is a geography which gives credence, life and agency to *animals themselves* rather than seeing them only through, and associated, with human frames of one kind of another? Donna Haraway's claim that "all ethical relating, within or between species, is knit from the silk-strong thread of *ongoing alertness* to otherness-in-relation" (2003, p. 50), offers the beginnings of an answer. More than a decade earlier Haraway stated that "biology and evolutionary theory over the last two centuries have reduced the line between humans and animals to a faint trace" (1992, p. 193). Since then, ongoing scientific, philosophical and creative revelations about animal life, and human life, have further questioned that dividing line.

Animals know and practise the world through spatialised embodiment, the world as place, territory, habitat, route, home, and so on. They *make* homes (e.g., nests, burrows, lodges, dens) and territories, and they do so in very complex relations with others of their kind and other species. This, again, is ecology. The stakes for animals are the very highest. For example, for migratory animals who travel thousands of miles to return to previously used and precisely remembered sites of breeding, feeding and home. These are not merely routine or instinctive practices, but skilled individual negotiations of the environment. Such animals face multiple spatial risks in the modern world, for example, loss of habitat, feeding and breeding opportunity in the lands that are their seasonal homes, and on the routes they travel between them. If human cultures fail to be hospitable to such travellers and such visitors, it is a very tragic situation and dangerous situation for all life.

The eminent geographer of space Doreen Massey (2005) insists that we take on the "challenges" and "coeval multiplicities" of space as this presents us with the social in the "widest sense" – "the radical contemporaneity of an ongoing multiplicity of others, human and nonhuman" (195). This seems especially so for how animal geographies suffuse through "social spaces" in many strange, beautiful, tragic, and often unregarded ways.

Now, as discussed in the chapters of this book, through various means and methods, we have the chance to pay the closest attention to animal life, and map and narrate animals' embodied, relational, spatial practices of becoming, and to set those alongside human becoming with a view to understanding and empathy through shared common ground. Fields of sciences such as zoology, ethology, ecology, zooarcheology, offer methods for considering animal life close up. New, interdisciplinary approaches which borrow from the social sciences, notably multispecies ethnography have great potential too. This paying close attention to lived animal lives can act as a counter to both social construction and other anthropocentric tendencies in modern knowledge and "objective" science which reject that which it cannot technically verify. This is a key point and has been, and remains, something of a front line between older and new approaches to animal studies. It is through a certain type of closeness, and being with animals habitually and over time, that we can come to know animal becoming and personhood. It is impossible to translate much of this, or to fully know, but it becomes clear from being with animals that they are "saturated with being" (Whatmore & Thorne, 2000, p.186). The eminent eco-philosopher Val Plumwood asserts that the behaviours of certain animals

> require sophisticated higher-order intentionality; there are so many examples of this kind, which so many people experience, that one has to wonder whether theorists who strive to dismiss them have any knowledge of animals outside the laboratory.
>
> (2002, p. 182)

This key point is returned to again in the chapter – particularly in the section on ending concerns about anthropomorphism.

Below, the chapter is organised into three more sections. The first sketches out the emotional and affective turn within geography and other subjects, and how this challenges human exceptionalism. A key theme throughout is that by paying careful attention to animal becoming in action we can engage with their affective lives and thus their personhood. The second section explores the idea that these new understandings of the continuities between human and animal life should put an end to concerns over anthropomorphism. The third section explores further how we might reveal and witness animal becoming and thus personhood through various forms of studying, witnessing and narrating. The last section focuses on forms of co-productive and participatory research with nonhumans which can be seen as an exciting new development to stand alongside, or extend, ideas of witnessing.

Affective Animal and Human Becoming

The affective "turn" in the social sciences and beyond, which emphasises the salience of relational, emplaced, material, non/pre-cognitive nature of human becoming, should ensure that no great distinction now remains between the qualities and values of human and nonhuman becoming. Affect radically challenges the idea of human exceptionalism and the negative connotations of anthropomorphism that have been used to belittle those who chose to talk about nonhumans in "human" terms.

Thinking with, and about, affect helps reveal animal lives and allow us to get close to them. Firstly, it highlights the animalness of humanity and the continuities between animals and humans. Secondly, it opens up this growing theoretical/methodological repertoire for considering animals themselves, albeit it that some methods, such as ethology, came from animal studies in the first place.

Affective registers are the systems and processes which make up much of life moment by moment, and which run pre, within and beyond reflexive, thought-based consciousness. These systems are in the body, and between the body and other bodies, and the body and the environment. These are the registers through which we live the bulk of our lives as beings-in-environment on an ongoing basis. Memory functions, emotions, motor movements (e.g., balance), sense/response systems are key but not exclusive parts of affect. As key scientific work (see, e.g., Damasio, 1999, 2003) has shown, our conscious, reflexive, rational, language using self (the focus of identity, conventional politics and ethics and much social science) is underpinned, enframed and constantly coloured by these processes. Affect is about the capacities of bodies to affect and be affected in relations with others and the environment (Boyd, 2017).

These insights into the biological, bodily basis of individual and collective social life challenge the Cartesian settlements of identity and agency, and the key dualisms of Modernism (as already discussed), mind/body, nature/

culture, subject/object. Affective life is very much about embodiment, *relationality* and materiality as it seeks to deal with the self as a performative entity always in specific space-time circumstances of the now (and legacies of past experiences). The social sciences which operate in and through representations, constructed in and of language and reflexive, rational, thought, thus struggle to "see" and work with affective life. This is why an expanded view of what constitutes meaningful life is needed, along with new methods which can work in the beyond-language world. It is important to note, at this stage, that affect and emotion are not the same things – but they are closely related and interact intimately.

The emphasis on affect/emotion is not to deny the power and importance of rationality, and cultural and economic dynamics within human-animal relations. These clearly shape human-animal interactions and their related spatial/ethical patterns, but affect and emotions will always be underpinning and co-present forces, and are processes through which the other dynamics are channelled or practised.

Concept Box 4.2: Affect, animal becoming and personhood

The "affective turn" has generated profound challenges to the social sciences and related disciplines in terms of what the social is and how it can be engaged with through ontology, epistemology and methodology. Below, a recent summary of affect, and why so many disciplines and scholars have directed their attention to affect:

> [A]ffect epitomizes a dimension of meaning in human affairs that is not a matter of established discourse, of stable identities, institutions, codified cultural norms or categories, but rather something that is lived, from moment to moment, at a level of sensuous bodily reality beyond codification, consolidation or "capture". Affect, on this perspective, is something that incessantly transgresses individual perspectives and frames of references (notably the perspective of the "autonomous subject" of the liberalist tradition). Affect is what unfolds "in-between" – in between interacting agents, in between actors and elements in communal everyday practices, within processes of transmission, be they medial, symbolic or aural, and in the involvement, absorption or immersion when the boundaries of the self become porous (or when they have not even been properly drawn to begin with). While it is impossible to grasp this sensuous immediacy directly, proponents of affect studies undertake it to cultivate a sensitivity for these fleeting moments, these shimmers, these stirrings of the nascent, the not-yet formed, the pre-reflective, the nuanced presences prior to reflection and articulation []. Such a sensitivity deviates from established

methodological canons and also, occasionally, from the strictures of theory. Practitioners of this strand of affect studies are accordingly inclined to explore poetic and personal styles, toy with allegiances to the arts, experiment with unusual modes of articulation and presentation.

(Slaby & Röttger-Rössler, 2018, pp. 1–2)

It is important to note that ecofeminist thinkers, such as Plumwood (2002), have long argued that we need to re-enter realms of feeling in order to build effective knowledge of human–nature–animal relations. Feelings, and other body-mind processes and capacities, are important parts of the overall idea of affect.

The key point is that the attention to affect in the study of human life shows how individuals are embodied beings in an ongoing becoming in place, with all the material and relational ecologies thereof. These processes fundamentally shape our experiences, our cultures and our identities. Much this process is, in turn, shaped by the capacities of the body and of the environment. We share many bodily and environmental capacities with animals, and thus we can assume that there is significant overlap in the experience of affective human becoming and affective animal becoming. The pay-off is then that studying the affective life of animals is a way of beginning to know their individual becomings, and through that, their personhood. There is a range of possible ways of doing this, but at heart, all these involve first watching animals very closely, over time, and taking the detail of their individual, emplaced embodied, relational lives as seriously as we take our own individual lives.

Ending Concerns about Anthropomorphism

Anthropomorphism is the attribution of human characteristics and emotions to nonhumans, which can range from objects (e.g., cars), to forces (e.g., wind) and, notably, to animals. There are long traditions in literature and folklore in giving animals human voices and emotions and placing them in stories. On occasion, this has had significant impacts on how animals are seen and treated in society. For example, the novel *Black Beauty* by Anna Sewell (1977), which told the life story of a horse and his companions in first person voice, as they moved through various regimes of human ownership and treatment in Victorian England, was said to have made an impact on attitudes to horse welfare in society.

But beyond literature and folklore, modern science has been suspicious of any form of anthropomorphism that sought to claim genuine insights into the emotional and affective lives on animals by assuming anything was sharable between the human and the animal. As we cannot objectively, and with certainty, get into animals' minds, we cannot know what they think and feel.

Thomas Nagel famously postulated that it was impossible for humans to successfully image what it is like to be a bat.

This was an attack on reductionist, materialist theories of mind, but Nagel did believe that conscious becoming was shared by many types of creatures, and that it is an innately subjective qualitative mode of becoming which can be acknowledged in general, but not known and shared in particular. In other words, we know they are alive, that they have some sense of that life, but we can't assume to know what that life is like. Even animal experiences of pain have been controversial, particularly in species with very different body and life worlds to humans, such as fish. But human pain is felt via the nervous system and brain, a system evolved to show when the body is sustaining damage, and to encourage the avoidance of such for purposes of survival. Fish have nervous systems and brains, their bodies can sustain damage, one would have thought the answer to can fish feel pain is really pretty obvious.

To take this way of thinking further here is a thought experiment. We, and in fact all living things, are cell-based organisms. Cells are highly complex things, but water is critical to their function, and thus to the body's function overall. In dehydration our cells start to shrink and malfunction. A few days without water we will really begin to suffer a decline, quite quickly, towards death. Our nervous system makes it very clear what is going on, we feel thirsty, and then deep levels of suffering from dehydration. Not only can we say that a dehydrating animal goes through the same body experience, and thus suffers, we can even think so for plants too. Their cells shrink in dehydration, their body, and becoming, is damaged and eventually killed by dehydration. Plants do not feel this in the way that animals do, as they do not have nervous systems. But what their cells, and body, are going through is the same process of damage. We know what that feels like. Tim Ingold insists:

> It is not 'anthropomorphic' [ ] to compare the animal to the human, any more than it is 'naturalistic' to compare the human to the animal, since in both cases the comparison points to a level on which the human and animal share a common existential status, namely as living beings or persons.
>
> (2000, p. 50)

He goes on to say that animals and humans are kinds of "organism-persons", making up "innumerable forms of animate being", they have in common the fact that "they are alive" (ibid).

The animal welfare theorist and policy pioneer John Webster has coined the term "reverse anthropomorphism" in an attempt to reconnect human and animal experience where, instead of thinking of animals with human qualities and experiences we turn the thinking around and think human as animals. Frans de Waal (1999) speaks of the "anthropodenial" that often prevents this happening, a blindness to the human-like characteristics of other animals or the animal-like characteristics of humans. De Waal, and others, notably Horowitz and Bekoff (2007), have used the idea of "informed anthropomorphism" as a means of stepping between polarised notions of anthropomorphism. Often in relation to considering relationships between humans and companion animals, this idea seeks to draw out the more obvious aspects of shared becoming in everyday interactions such as play, touching and cuddling between, for example, humans and their dog companions.

We are aware, to some extent, of our own experiences of embodiment, of pain and pleasure and everything else. We know that much! So we should extend it to the animals who share so much common ground with us. The Cartesian border controls that our complaints about anthropocentrism become illegitimate. If there are continuities of form and function, there can be continuities of other kinds. If our feelings, affective processes, interactions, language are sophisticated versions of that which occurs in animals then there will be much common ground. Philo and Wilbert (2000) suggest that we "should extend human 'courtesies' to animals", and this process may stem from a "certain anthropomorphism (reflecting the possibility that in certain respects animals are not so different from humans)" (25).

The common ground between humans and animals (and nonhumans of differing types) is not, cannot be, language in our sense of it. The common ground is affective embodiment and regard for and of it. We (modern humans) have come to think and speak in words and sentences. Language has become our voice. Much of our affective becoming, the unconscious, the body, movement, sense, balance, memory function has been relegated to a level "below" voice. But we don't sense, emote, or move in words and sentences, all those things we do share, to differing extents, with animals. Human language and higher consciousness may be forms of blindness to other ways of knowing the world, to other ways of becoming which are common in animals. Animals' "knowing", and becoming, may work in the different registers of senses, bodies and practices-in-space. The suggestion is that we can know something of them, and give voice to their becoming, by paying careful attention to that becoming in action.

Assuming sharedness through anthropomorphism, if that is what we have to call it, seems not to be problematic, but a good way of thinking at least certain animals. Can we read the body movements, and body moments, of animals relatively unproblematically? I would say we often can because they are shared embodied processes between humans and many similar animals. I watch our cats yawn and stretch after sleeping or skitter around the garden in a state of animation, or squirm luxuriantly in the sun. The bodily, affective similarities between these moments of becoming and human yawning and stretching, human skittering, and so on, means there are profound likeness between them which are, for that moment, as strong as the obvious differences which also persist – both are in very distinctive similar modes of affective bodily becoming. There are many other states we share with animals – becoming-scared, becoming-content, becoming-tired; these are states of affect, or perhaps waves of affect, and always articulated in the body, through posture, movement, stillness, and in relational material, emplaced, entanglements. As for animals which are much more different from humans in kind, such as insects, fish, reptiles, shared body-becoming might be harder to imagine. But in response to that challenge, I suggest that, firstly, the ground of shared life is a starting point. A body is a body. A body in place is a body in place; whatever kind it is. Secondly, we need to heed those who study kinds of animals in great detail and often, with great love. For example, listen to the TED interview with Sylvia Earle (Earle, 2019) about her life as an international pioneering marine biologist, and when she tells specific stories of witnessing fish sentience, and individual fish personhood, and maybe you will think, OK, she has looked, she has seen, she is a witness to animal becoming and personhood. A quote from this interview appears later in the chapter.

Revealing Animal Life through Watching Closely, Witnessing and Narrating

We now move on to consider in more detail the revealing of animal life. In various ways, people can act as loving-expert witnesses, as advocates, as spokespeople for nonhumans and their personhood, as Earle does for fish life. Getting close to nonhumans, and observing them attentively, can be done through various lenses, for example, ethology, multispecies ethnography, philosophy, literature, popular natural histories, and artistic practice, or simply living attentively and empathetically with animals at home and in the neighbourhood.

Animal lives can be watched, witnessed and *narrated* in several ways. Narratives can become a common ground between us (all), a bridge between the human and nonhuman, a means of positive anthropomorphism, a way of making all humans and nonhumans a throng of differentiated, significant others in which the nature/culture divide dissolves. Witnessing can be developed as in-depth, committed, ethical and empathetic watching and narrating.

If we regard animals closely, *over time*, their spatialised becoming-withs can begin to come to life for us. To return to cats as an example, if you have the opportunity, watch a cat while he/she is alert, or even half asleep, and watch the ears move as they track various sounds. Cats can move their ears to focus on some sound or other in ways similar to how our eyes move, listening intensely to one source. But then another sound elsewhere might drag one ear off briefly to point in another direction. Cats can listen to two sounds from different places at the same time. Imagine if human eyes could do that! I have watched one of our cats, on a stormy (wind noise) day – with people moving around the house at the same time, with her ears twitching one way, then another, sometimes together sometimes individually. What a complex mental process it must be to actively listen to two different things at once (like our eyes deliberately looking in two separate directions at once). Try to begin to imagine such different practice of (aural) space. Cats' worlds and the personhood they construct within them are sound worlds much more than ours. The cat in question is in a unique moment of space and time which is shaping her becoming on an individual basis.

In Coetzee's novel *Elizabeth Costello*, the eponymous lead speaks of poetry which tries to deal with animal embodiment: "With [Ted] Hughes it is a matter – I emphasize – not of inhabiting another mind but *of inhabiting another body*" (2003, p. 96). "In these poems we know the jaguar not from what he seems but from the way he moves. The body is as body moves, or as the currents of life move within it" (pp. 95–6, emphasis added). What is forming here is "voicing" animals through their ways of becoming rather than our (human) ways of becoming. Ted Hughes is rightly famous for his poems about animals, both wild animals and farm animals, which focus very intensely on not only their uniqueness as species, but on their embodied presences as individual living beings.

In human-animal studies, Peterson (2001) asks us to be mindful of "the worldviews of other species – their knowledge of a given landscape, their own construction of reality. [...] We ought to ask whether there are nonhuman ways of being conscious, of having a mind" (p. 223). Because we cannot fully know animals, this does not mean to say we have to assume them void of mind, any more than we do human others. Many people with interests in animals have made this kind of point.

Kay Milton (2002, p. 50) is confident that her cat can tell her "I'm hungry" as well as any person can – through relational body/voice language exchanges within the closeness, intimacy and trust of their ongoing life together. Haraway (2008) similarly tells of a "determined" dog, insisting (by putting a paw on her keyboard) that it is time for their daily walk. As Anne Game (2001) summarises, "people who live with animals experience connectedness and cross-species communication daily" (pp. 1–2).

From the world of "alternative agriculture" the farmer and writer Rosalind Young in her celebrated book *The Secret Life of Cows* (2003/17) describes in detail the natural history of the cows on her farm that are left to behave as naturally as possible on her farm (UK) while still being productive in agricultural terms. She tells stories of how individual cows relate to and exploit the environment in terms of shelter and food, and how they socialise with each other and with humans. She tells *individual* stories and *family* stories, about how some cows strike up friendships, while others steer well clear of each other. She tells how differing cows have markedly differing mothering practices and skills, and describes what she calls "signs of intelligence" such as young calves learning tactics for accessing food at crowded feeding troughs where they are physically out-muscled. All this is done by watching and recording the cows' movements in terms of where they choose to go, what they do, who with, and their practices and their body language. Not only are there obvious welfare issues at stake here, but matters of efficiency. Her methods result in very healthy and high-quality animals in terms of commodity production. Young does not profess "becoming-cow" or to any particular and specialist scientific approach. Her method is "simply" careful, patient watching of the cows over a long period of time.

Haraway (2008) refers to another "radical" UK livestock farmer, Rowell, who approaches rearing a flock of sheep in similar ways. Much can be learnt from new developments in animal behaviour and (farm) animal welfare studies. We can also watch closely the animals around us – our companion animals, and those we might see in landscapes around us. There are implications for "practical" questions of farm animal welfare, such as "the five freedoms" (a code of practice developed to underpin farm animal welfare), and yet further questions about ethics and animal (non)becoming when we begin to seek out animal geographies in this way.

MacFarlane (2005, p. xii) in his introduction to one of the most famous accounts of intense wild animal watching, J.A. Baker's *The Peregrine* (2005), says it "is a book in which very little happens, over and over again. The man watches, the bird hunts, the bird kills, the bird feeds". What Baker does, apart from struggling to find the peregrine(s) in the landscape as they live quite secretive (to human) lives, is to describe again and again, in differing essays of language, the bird's embodied life, its flight, stoop (hunting dive), feeding, roosting, its directions, and daily and seasonal rhythms. And what emerges are *the repertoires* of all these, and these repertoires in contexts of space, season, weather, light, local landscape, other bird life and Baker himself.

The bird's flight varies between light and flickering, heavy and laboured, fast and purposeful, slow and indecisive, straight or arcing, high or low (my terms). It is tempting to link these differing bodily expressions with the bird's inner mood – its becoming self – but what we should say is that they *are its mood* – a mood of movement, linked, no doubt, to affective body processing (tiredness, hunger and excitement).

Kringelbach (2009) shows that the parts of the brain that process movement are linked to those that process emotion. Moods are embodied as in felt in the body, (e.g., say, fear which cause muscles to tense), as well as atmospheres of the mind, and are thus readable across the human/animal divide.

Other more recent literary treatments of avian becoming such as Mark Cocker's *Crow Country* (2007) also spend many a page describing the differing individual and collective flying movements, calling repertoires and nesting/feeding spatial patternings of corvid colonies. As in J. A. Baker's *The Peregrine* (2005), the birds' becomings are filled out by fine-grained narration of their relational, spatial, material embodied everyday lives. In other words, their living geographies. These in turn can only be garnered by dedicated hours, days and even years of close watching.

If we pay close attention to the detailed practices of animals' spatial becomings and render them into readable narratives, we begin to shift our telling of them onto their ground. But it is not easy. We (humans) are perhaps inevitably bound into understandings of becoming, self and personhood through processes of reflexive thought, language and identity. As Ingold (2000, p. 49) has put it:

Western thought [ ] drives an absolute division between the contrary conditions of humanity and animality, a division that is aligned with a series of others such as between subject and object, persons and things, morality and physicality reason and instinct and, above all, society and nature.

In the end, advice could be to just witness whatever individual animals you encounter in your everyday life, be it pets in the house, city animals such as pigeons, seagulls, foxes, farm animals, or working or recreational animals such as horses. This can even apply to insects and other forms of animals such as spiders and beetles. To watch a spider making a web, as I have filmed at home, is to see someone absorbed in their work, in their world making, with calm methodical movements. To watch a spider, trying to climb out of an empty bath, having come down looking for water at the fringe of the plughole, is to see someone scrabbling in panic and confusion as they find the climb hard, or impossible.

The philosopher Raimond Gaita's (2003) contemplations of animals, and the nature of minds (both human and nonhuman minds) are, in large part, based upon observations of those animals he has lived with, such as his dog Gypsy, and Jack, his father's cockatoo. This is significant because it is through this closeness, this ability to attend to animal becoming over time, watching them closely, in a relationship of trust and familiarity, that we begin to know them and maybe can speak of/for them. Gaita also draws upon the writings on animals by Coetzee and the philosophy of Wittgenstein. In a key phrase he suggests that "an animal's intelligence is entirely active, its understanding entirely practical" (p. 61).

Their becoming resides in the practices which, in humans, are relegated, but which, in actuality, still form and frame us. Damasio (1999), whose work has been critical in developing geographies of affect and emotion, has shown that affect, emotion and feeling underpin, initiate and allow thought episodes, and not vice versa. These processes form the vast bulk of our becoming; they are, perhaps, the entirety of animal becoming "it is clear that many nonhumans have emotions in abundance" (Damasio, 1999, p. 35). How then is that lived and expressed and studied?

I make two final brief points in this section which pertain to very practical issues of human and animal wellbeing in relation to affect. Firstly, it is now well established that in certain circumstances the access to animals for humans who are ill can be very beneficial in reducing stress symptoms. Dogs, with their very special affinities with humans, are used, for example, in old people's homes and some hospitals as companion animals to watch, stroke and talk to. I suggest that what is going on here is the sharing of affective becoming between human and animal. The (trained) dogs in question are happy to be in the situation they are in, satisfied at some deep level of personhood. Those humans who come into contact with that through observation and touch and sound, pick up and share that affective resonance.

The second point, going right back to an opening point about mobilising performativity instead of "inner" emotions, is that emerging and innovative (and to some extent controversial in conventional science terms) animal welfare assessment protocols are being developed where careful attention to an animal's movements and "body language" are in fact very reliable ways of assessed animal welfare and physical *and* mental health (see Wemelsfelder, 2007).

Co-Creating Narratives of Nonhuman Spatial Becoming Through Participatory Methods

Recent developments in participatory research methods with nonhumans (see, e.g., Bastian et al., 2017) offer an exciting but challenging way of thinking about how we might develop narratives of animal becoming – beyond narratives by humans about animals, to narratives created *with* animals. Of course, there is not an overly sharp distinction or dichotomy here but a blurry cross-over space. Empathetic stories of animal becoming, created by careful witnessing, could be said to be made with animals; for example, by spending time with them.

Time has been a theme through this chapter, as, quite simply, time is needed to get to know someone. Time is needed to get to know animals, and individual animals. This point was beautifully made in the aforementioned interview with the marine biologist Sylvia Earle in relation to sea life and fish. She recounted how, from a very early age, she swam in the sea, dived in the sea, and watched the wonders of marine life. Once she had become a marine biologist, she said that technologies such as scuba diving gear and submersible observation vehicles gave researchers, "the gift of time", being able to see these worlds and the creatures therein on an ongoing basis.

> To get to know individual fish, to recognise their faces, to see their individual behaviours, to see how that group of angel fish stick together, like a group of buddies swimming around together by day, to see the butterfly fish, that actually mate for life, like some people; look at the barracuda, not just as "ohhh, barracuda!" but *that* barracuda, different from that one over there. Some are more curious, others are more reserved [ ]; or the sharks, they all have faces. They all have personality, and, as some silver-sided scientists might say, as they did to Jane Goodall, "why, you can't look at those chimpanzees as individuals, you can't give them names, that is unscientific". But it is just a fact, they *are* different. They do have personalities. You might as well face up to it. And use that as part of understanding; who they are, who we are, and how we are part of a single, interacting system that we call life on earth. [ ] People say, I have heard this so many times, fish don't feel pain, they don't have the intellectual capacity to feel pain. Well, excuse me, put a hook in your mouth, and what would you do? You would do exactly what a fish does. Of course they feel pain. I say it is intellectually inconceivable that you would think otherwise, knowing that they are fellow vertebrates. They have a backbone, we have a backbone, they have a brain, we have a brain.
>
> (Earle, 2019)

Pushing the idea of participatory input by animals in the construction of narratives of them could play a part in making the earth more just, more co-liveable, more co-flourishing for humans and nonhumans than it is at present. The participatory, co-created, research movement to look into human life is now well established and has carried out a great deal of progressive work in a range of social settings. As is discussed in Bastian et al.'s edited collection on participatory research with nonhumans (2017), free-to-access methodologies, tool kits and ethical protocols are in place as resources for those seeking to conduct such work (see, e.g., Pain et al., 2011), and for further refinement in methodological terms for nonhumans.

Participatory research seeks to fundamentally change the way in which knowledge is produced. Instead of scholars choosing a subject of study, studying them, and then analysing the gathered data and to come to conclusions, participatory methods seek to dissolve the distance and asymmetry between those being studied and those doing the studying. In participatory, co-created research, a dynamic is created in which both "sides" move to new positions of attitude and action. Those who would have been, in conventional methods, the objects of research, now move to creatively inputting into the process, and those who would have been designing and carrying the research, embrace sets of negotiations and discussions which change the objectives, the methods and potentially the whole process from beginning to end.

Participatory methods in the human realm are based on the flattening of power relation differentials; conversation cycles, developing trust, changing the temporal structures of the research process, ensuring common languages, vocabularies and terms are developed, ensuring that affective arrangements are adjusted to allow for genuine co-creation. Such procedures must be significantly adapted to bring animals and other nonhumans into the process. To do participatory research with nonhumans raises a whole suite of conceptual, ethical and practical challenges. Most obviously, language poses a significant challenge. Participatory research is based upon structured conversations, agreed agenda, cycles of action and reflection. Power relations in the process pose another significant challenge. In participatory research all participants are, as far as is possible, put in an equal position of power. To do all this with animals, particularly those more other in life form than humans (insects, fish and so on), demands imaginative and empathetic procedures to find other ways of doing things. At this point, the focus must be on individual animals in individual situations, because all situations are different. A whole range of scholarly and creative work across the arts, humanities and social sciences can bring specific disciplinary research methods to do this.

We need to develop the excellent work already conducted under the auspices of more than human research, animal/nonhuman geography (and related subject areas) by stressing various aspects of the affective, spatialised, embodied becomings of nonhumans and seeking ways of their actively speaking into knowledge creation. As Bastian (2015) outlines:

> Widespread interest in challenging the traditional divides between humans and nonhumans has contributed to a growing push for methods that can work with the distributed knowledges, experiences and values of

our multi-species worlds. In response, proposals for the development of etho-ethnology and ethno-ethology (Lestel et al., 2006), multi-species ethnography (Kirksey & Helmreich, 2010) and zoömusicology (Taylor, 2013), amongst others, have augmented, hybridised and remade method-ological repertoires.

To return to debates within animal geography Philo (2005), in reviewing Whatmore's (2002) notable contribution to animal/nonhuman geography, suggests that her work focuses on the performativities and material relation-ality of animals (their observable practices) rather than trying to imagine their interior life as expressed in thoughts, feelings, intentions and emotions. He asks whether, even in these attentive accounts of animal life, the individ-ual animal remains ill-formed as a significant being:

> And yet, might it not be that [the] animals - in detail, up close, face-to-face, as it were – still remain somewhat shadowy presences? They *are* animating the story being told [], but in their individuality – as different species, even as individuals – they stay in the margins.
>
> (p. 829)

This then might be the next step for all manner of animal studies. My pro-posal, in some ways bridging between these two positions, is that animals express their souls, their being, their personhood through individual bodily, relational, spatial practices (and therefore in sensible terms), so paying close attention can bring them into focus as meaningful individual beings.

Conclusion: Continuities, Discontinuities and Empathies Between Humans and Animals

In this chapter I made a series of interlinking arguments as follows: that ani-mal becoming is inherently embodied and inherently spatial, and in that indi-vidual embodied spatial becoming is their personhood and meaningful life as living beings. Of course, it should be added that really all becoming is spa-tio-temporal, so we need to think through that double lens about animal becoming, animal personhood and how much human-nonhuman becoming shares registers, emotions and affects. In short, all living things share life through space and time. Yes, there are many profound differences, but the sharedness is more fundamental, and foundational, than those differences. We can thus assume that our ideas of human personhood, should be shared with animals and, somehow, probably with all living beings.

Many processes of life, both animal and human becoming, and becom-ings-with, occur through affective registers. As human and animals share bodily capacities, and also live in the same material world (albeit differently), there is plenty of shared affective becoming between humans and animals. We can not only assume they have personhood, we can know some shared aspects of it.

The surge in interest in affect in both human and animal studies shows us that many bodily states of becoming are shared between humans and animals. This then dissolves the boundary between them erected by Modernity. Affectively sensitive approaches can thus challenge human exceptionalism, bring humans and animals together in many ways, and challenge the anthropomorphism used to argue against empathy and care for animals. All this can feed into senses of creative, collective ecological selfhood, and new practices of community, place and earth.

A positive notion of anthropomorphism would not simply suggest that animals are like humans, it would suggest that there is *plenty* of shared ground between humans and many animals, and that there are continuums of spatial, relational bodily practice through all animal kind, and a continuum of embodied relational life between all living beings.

As already stated in the discussion about participatory research, various disciplines and practices in the natural and social sciences, and the arts and humanities, can begin to meaningfully engage with animal personhood though close readings and witnessing of their spatial becomings. The development of participatory and co-created research with animals holds out much promise in this regard.

In conclusion, the shared ground of life is *life itself* (between all living things); then, *life as mobile, embodied, emplaced becoming* and then, becoming-with, how all live in relation to all. That is a fundamental ground which means that humans and animals/nature are not distinct in kind. Beyond that, in many instances, quite specific shared physiological and affective processes of embodied becoming exist. This is the ground for both human and animal personhood. Furthermore, as discussed in relation to water and the body, it shows that all life has fundamental, shared forms of becoming. Once animal personhood is acknowledged, notions of animal cultures (which modern thought has also denied) soon follow.

Modern systems of thought, and the ethics, economies, politics and cultures they underpin have rendered us blind to the personhood of animals, to the values of animals beyond the functions of economy, and created human exceptionalism which is, mostly, an illusion. That is a key driver of the very grave planetary crisis we are now in the midst of. We are, I hope, on the long road to some radically new formulations of human-nature/animal understandings and practices that will underpin future society. This chapter, and book, are small steps along the way.

Acknowledgements

I am greatly indebted to the animals I have lived with on the farm and my long-term companion cats. I am also indebted to the editors of this volume for the extensive support, patience, and very insightful comments which have gone into the development of this chapter. Thanks also to Wendy Knepper and Professor Clara Mancini (Open University UK) for reading the draft text and helpful comments.

Note

1 See Kirk (2014) for a summary of intersubjectivity when training dogs to detect mines.

References

Al O'Kane (2015) Animals. *Al O'Kane*, first track. https://open.spotify.com/album/71FEoO3Pp7zJ0JE045XV3k?highlight=spotify:track:6bbSeY3UrVaurw4BeiFsZl

Baker J. A. (2005) *The Peregrine*, New York: New York Review Books.

Bastian M. (2015) *Multispecies Methods: Refractions of participatory research with more-than-human approaches.* Human Geography Seminar Series, University of Reading. https://www.reading.ac.uk/climate-justice/ltcj-events.aspx (accessed 31 01 2020)

Bastian M., Jones O., Moore N., and Roe E. (Eds) (2017) *Participatory Research in More-than-Human Worlds*, London: Routledge.

Boyd C. (2017) *Non-Representational Geographies of Therapeutic Art Making Thinking Through Practice*, London: Palgrave MacMillan.

Casey E. (1997) *The Fate of Place. A Philosophical History*, Berkley: University of California Press.

Cocker M. (2007) *Crow Country*, London: Jonathan Cape.

Coetzee J. M. (2003) *Elizabeth Costello*, London: QPD.

Damasio A. (1999) *The Feeling of What Happens: Body Emotion and the Making Consciousness*, London: William Heinemann.

Damasio A. (2003) Mental self: The person within. *Nature*, 423(6937), 227–227.

Darwin C. (1872) *The Expression of the Emotions in Man and Animals*, London: John Murray.

Despret V. (2004) The body we care for: Figures of anthropo-zoo-genesis. *Body and Society*, 10, 111–134.

Earle S. (2019) *Sylvia Earle makes a case for our oceans*. A TED Original Podcast. https://www.ted.com/talks/the_ted_interview_sylvia_earle_makes_a_passionate_case_for_our_oceans#t-942058 (accessed 26.09.2019)

Frans de Waal B. M. (1999) Anthropomorphism and anthropodenial. *Philosophical Topics*, 27(1), 255–280.

Gaita R. (2003) *The Philosopher's Dog*, London: Routledge.

Haraway D. (1992) A Manifesto for Cyborgs: Science. Technology, and Social Feminism in the 1980s, in L. J. Nicholson (Ed) *Feminism/Postmodernism*, London: Routledge.

Haraway D. (2003) *The Companion Species Manifesto. Dogs, People, and Significant Otherness*, Chicago: Prickly Paradigm Press.

Haraway D. (2008) *When Species Meet*, Minneapolis and London: University of Minnesota Press.

Horowitz A. C. and Bekoff, M. (2007) Naturalizing anthropomorphism: Behavioral prompts to our humanizing of animals. *Anthrozoös*, 20(1), 23–35.

Ingold T. (2000) *The Perception of the Environment. Essays in Livelihood, Dwelling and Skill*, London: Routledge.

Kirk R. G. (2014) In dogs we trust? Intersubjectivity, response-able relations, and the making of mine detector dogs. *Journal of the History of the Behavioral Sciences*, 50(1), 1–36.

Kirksey S.E. and Helmreich S, (2010) The emergence of multispecies ethnography. *Cultural Anthropology*, 25 (4), 545–576.

Kringelbach M. L. (2009) *The Pleasure Center: Trust Your Animal Instincts*, New York: Oxford University Press.

Lestel D., Brunois F., and Gaunet F. (2006) Etho-ethnology and ethnoethology. *Social Science Information*, 45(2), 155–177.

MacFarlane R. (2005) Introduction to J. A. Baker. *The Peregrine*, New York: New York Review Books, vii–xv.

Massey D. (2005) *For Space*, London: Sage.

Milton K. (2002) *Loving Nature: Towards and Ecology of Emotion*, London: Routledge.

Pain R., Milledge D., and Lune Rivers Trust (2011) *Research Toolkit: An Introduction to Using PAR as an Approach to Learning, Research and Action*, Durham: Department of Geography, Durham University.

Peterson A. L. (2001) *Being Human: Ethics, Environment, and Our Place in the World*, Berkeley: University of California Press.

Philo P. (2005) Spacing lives and lively spaces: Partial remarks on Sarah Whatmore's hybrid geographies. *Antipode*, 37(4), 824–833.

Philo C. and Wilbert C. (2000) Animal spaces, beastly places: An introduction, in C. Philo and C. Wilbert (Eds.) *Animal Spaces, Beastly Spaces: New Geographies of Human-Animal Relations*, London: Routledge, 1–34.

Plumwood V. (2002) *Environmental Culture: The Ecological Crisis of Reason*, London: Routledge.

Rowlands M. (2019) *Can Animals be Persons?* Oxford: Oxford University Press.

Silvey R. (2017) Bodies and Embodiment, in D. Richardson, N. Castree, M. F. Goodchild, A. Kobayashi, W. Liu, Richard A. Marston (Eds) *International Encyclopedia of Geography: People, the Earth, Environment and Technology*, London: John Wiley & Sons, 370–376.

Singer P. (1975) *Animal Liberation: A New Ethics for our Treatment of Animals*, New York: New York Review/Random House.

Slaby J. and Röttger-Rössler B. (2018) Introduction: Affect in Relation, in B. Röttger-Rössler & J. Slaby (Eds) *Affect in Relation – Families, Places, Technologies. Essays on Affectivity and Subject Formation in the 21st Century*, New York: Routledge.

Smuts, B. (2001) Encounters with animal minds. *Journal of Consciousness Studies*, 8, 293–309.

Taylor H. (2013) Connecting interdisciplinary dots: songbirds. 'white rats· and human exceptionalism. *Social Science Information*, 52(2), 287–306.

Weber A. (2019) *Enlivenment. Towards a Poetics for the Anthropocene*, Cambridge MA: MIT Press.

Wemelsfelder F. (2007) How animals communicate quality of life: The qualitative assessment of animal behaviour. *Animal Welfare*, 16(S), 25–31.

Whatmore S. (2002) *Hybrid Geographies: Natures, Cultures, Spaces*, London: Sage.

Whatmore S. and Thorne L. (2000) Elephants on the move: Spatial formations of wildlife exchange. *Environment and Planning D: Society and Space*, 18, 185–203.

White L. (1978) *Medieval Religion and Technology*, Berkley: University of California Press.

Young R. (2003) *The Secret Life of Cows*, Preston: Farming Books and Videos.

Part II
Collaborating

5 Two Species Ethnography

Honey Bees as a Case Study of an Interdisciplinary "More-than-Human" Method

Siobhan Maderson and Emily Elsner-Adams

Introduction

The 21st century has seen a realisation that the world we inhabit has exceeded human control and representation (Dowling et al., 2017), with humans and nonhumans facing unprecedented threats (Taylor & Twine, 2014). Problems such as climate change, coupled with a better understanding of how ecosystems underpin and support human lives (Millennium Ecosystem Assessment, 2005; IPBES, 2019), are encouraging new approaches to engaging with the world, both through spanning disciplinary boundaries and by re-forming the assumptions and attitudes we hold about our nonhuman cohabitants of Planet Earth.

Animals have been part of human lives for millennia, as food sources, objects of worship and spiritual kinship, companions and recently highlighted as *supporting* human existence through ecosystem services. While many peoples have never ceased to acknowledge their interspecies entanglements, Western post-enlightenment worldviews supported a division between species, resulting in practical and philosophical reflections on how we relate to other species. Although primarily focused on mammals, animal rights have been discussed across diverse taxa, even including insects (Lockwood, 1987). However, society in general, including the social sciences and Human-Animal Studies (HAS), tends to prioritise and focus our attentions on warm-blooded animals above insects, fish and other species that inhabit very different habitat spaces or lifeways to humans (Bear & Eden, 2011). Caught up in the generic noun "animal", it is easy to talk about nonhumans in abstract and collective categories and to ignore their lived experience as individuals, and as communities, entangled with humans, with their environment and with each other.

As we aim to expand HAS beyond the more easily discernible and anthropomorphised companion species of dogs and horses, any efforts to grasp the lives of less charismatic species such as fish and invertebrates will require methods which engage with human intermediaries, proxies and indirect sources of information. Some may see the reliance on human intermediaries as problematic since the field of HAS understands giving voice to other

DOI: 10.4324/9781351018623-7

species as its central raison d'être. We argue that the sheer scale of other species, such as fish and invertebrates, with which humans share the planet, necessitates a breadth and flexibility of method if we are to seek to do justice to deepening our understanding of the rich diversity of cohabitants on Earth. While efforts to engage more with Indigenous Peoples and Local Communities (IPLC) are part of this shift (Hill et al., 2020), for those scholars working primarily in the Western ontological tradition, and HAS, other methods may assist in bridging world views. In this chapter we present a methodology, Two Species Ethnography, that addresses the need for temporal contextualisation of animal and human behaviours in a rapidly, anthropogenically, changing world.

This chapter draws on our experiences of engaging closely with honey bees[1] (*Apis mellifera*) and beekeepers. Honey bees live a life deeply entangled with humans – particularly beekeepers. Working with this entanglement led us to develop an interdisciplinary, mixed-methods approach that we call "Two Species Ethnography" (TSE). TSE is a method for people who are working in an applied, case study context, focusing on a particular nonhuman species. While TSE acknowledges the wider theoretical aims and objectives of multispecies ethnography (MSE), TSE strives to be an accessible method for social and natural scientists aiming to understand a particular species. Doing so often requires engagement with two species: the particular nonhuman species in question, plus the humans that co-exist and interact with it. TSE provides conceptual and practical boundaries that some students may find lacking in MSE. It also reflects concerns among some political ecologists of the need to ensure attention is given to material realities, as well as conceptual understandings of the forces that affect the multiple species that inhabit our world (Lave, 2014). Many animal species are initially or primarily understood via the natural sciences, and/or particular groups of humans who engage with them via hobbies or professional practices. TSE recognises the need for HAS researchers to engage across academic disciplines and with human intermediaries in understanding some specific, often less charismatic[2] species. We note that there is (often extensive) natural science information available about nonhumans, as well as specific methods developed to observe, and ultimately understand the behaviours, life-history, ecological interactions and many other physical and mental aspects of nonhumans. This work comes from ethology, biology, zoology, micro-biology, medicine and many other disciplines. However, to understand the relationships between nonhuman animals and their human collaborators requires much more than just "doing science". In TSE practice, human intermediaries are understood to translate the lived experience of nonhumans into information that is tangible for other humans, and thus researchable by HAS scholars interested in a particular species. We also note the importance of observations and knowledge generated by a wide range of human stakeholders, who are highly attuned to the natural environment and particular species.

We suggest that there is an opportunity for the social sciences, specifically HAS scholars and others engaged in the "animal turn", to actively attend to both natural science literature, and relevant observations of other human intermediaries, like beekeepers. While the language and methods may be alien to many social scientists trained in working with humans, drawing natural science research into a conversation with work throughout the social sciences, including HAS, animal geographies, more-than-human geography, multispecies anthropology, sociology and more will result in a more nuanced understanding of the particular animal species and/or population being studied. In TSE, we call for deeper engagement with the substantial and diverse information about nonhumans generated via both formal scientific and tacit methods. All of these methods, and the information gleaned by them, are valuable for those studying human-animal relations, as we show later in this chapter. This is the challenge for contemporary HAS scholars to draw this literature into conversation with the well-understood work in HAS, and thus bridge the historic divide between the social and natural sciences. By doing so, we are better placed to elucidate the reality of the anthropogenic world which humans and nonhumans co-inhabit.

We start by briefly contextualising HAS in the history of other relevant disciplines, to understand how its development as a field mirrors the progression of other fields. Then we explore TSE as a method more closely, focusing first on the ethnographic and human research elements, and then on the more biological and ecological elements. While there has been a considerable amount of theoretical development in methods for working with nonhumans and humans together, various authors have noted that the implementation of these in practice is still lacking (Cudworth, 2016; Dowling et al., 2017). Therefore, we aim to make this section as practical as possible to aid students and other researchers in implementing HAS in practice. To this end, throughout this section, we use the case study of honey bees to illustrate the advantages and challenges of a TSE approach and why taking such an approach can enrich a study of human-animal relationships. We conclude with a brief discussion of applying TSE to other species and in other contexts.

Contextualising Human-Animal Studies

HAS emerged from concerns around human exploitation of nonhuman mammals in the 1960s and 1970s (Taylor & Twine, 2014). To date, social scientists have excelled in investigating the human aspects of HAS for a few animal species (primarily mammals, especially companion and farm animals), but researching animals more broadly and tapping into the lived experiences of species which humans may rarely encounter, such as wild animals, or non-mammals who we feel less immediate empathy with, such as fish and insects, continues to be a challenge (Hodgetts & Lorimer, 2015).

Synchronous to HAS' development has been the growth of the international environmental movement, whose diverse stakeholders seek to address anthropogenic ecosystem degradation and species and habitat loss. At its core, the environmental movement is about the ideological and ethical engagement of humans with the environment (the chemical, biological and other systems that create "life" on Earth, as well as the species and species collectives that comprise habitats). Hard or soft, the scientific approach seeks to explain a species from both the inside out (its biochemistry, chemistry, morphology, genealogy) and from the outside in (its behavioural drivers, environmental and ecological relations, and its inter- and intra-specific relations). This mirrors the interests of HAS, as outlined by Cudworth (2016), but draws on a completely different framing of animals, considering them to be objects, or mechanistic beings responding to stimuli (Lestel, 2014). Despite this, it is the animal behaviourists and biologists who have sought, through their observational and mechanistic methods, to make wild animals less unpredictable (Buller, 2014), especially those nonhumans that are deeply dissimilar to us. Through direct observation and experimental manipulation of individuals, groups and their habitats by ethologists, ecologists, zoologists and others, the activities of nonhuman species can be elucidated.

Critiques of the environmental sector reference the fact that large, charismatic mammals (Lorimer, 2007) often dominate funding and public attention, while perhaps as important yet smaller, less charismatic animals may be excluded. This exclusion is particularly pertinent in the context of non-mammal species that play an important role in ecological systems, but whose activities and life complexities may be hard to observe directly. Furthermore, some natural science research processes, through a close focus on an animal's internal and behavioural elements, can isolate the individual animals from their ecological context and from their interactions with humans. TSE as a methodology aims to resituate specific animal species within their ecological and human context, but also expands ideas of both what animal species can be studied and what is considered expertise beyond the natural sciences.

Of interest in the TSE context is that the perception of animals as objects deeply influenced the development of the discipline of ethology, the science of animal behaviour. Early ethology has since been critiqued for ignoring the emotional aspects of animal lives (what animals themselves feel, perceive, and know in relation to their own behaviour). This was recognised as limited and unacceptable in modern ethology (Bussolini, 2013) and was followed by the development of the sub-discipline of cognitive ethology (Jensen, 2017). New research, notably from cognitive science, as well as animal geography and sociology, has challenged historic perceptions of the uniqueness of human traits like language and emotion, and demonstrates that many nonhuman species participate in complex thought, forward planning and communication (albeit a-lingual communication) (Bussolini, 2013; Crist, 2004).

Concept Box 5.1: Ethology

Ethology is the study of animal behaviour. Classical ethology was established as a formal academic discipline in the 1940s (Griffiths, 2007) but has its roots in the work of natural historians in the 1900s and earlier, drawing on their detailed observations and illustrations to develop more formal understandings of animal behaviour (Griffiths, 2007; Bolduc, 2012).

Ethology draws on two trends within natural history: first, the desire to describe and investigate individual animals for their own sake (thus focusing on taxonomy and morphology); and second, the desire to draw connections between facts and observations about the natural world, and sort these into a system or relationship (Bolduc, 2012). Early ethology focused on mechanistic processes to try and explain why animals behave as they do – an approach that mimicked, and was influenced by, the taxonomic tendency to compare animals as discrete units of features (Bolduc, 2012). Ethologists understood behaviour itself to be an evolved process, determined by an organism's genotype, unlike comparative psychologists who understood behaviour as influenced by development and experience. These opposing views were a major intellectual conflict within 20th century biological sciences (Greenberg, 2012).

As ethology developed, it shifted from so-called "cabinet naturalists" to naturalists who studied living animals and carried out fieldwork (albeit often on animals in captivity). Some of the early and highly influential ethologists, namely Nikolaas Tinbergen and Konrad Lorenz, drew on earlier evidence showing that observation of zoo animals could provide valuable behavioural data. Their breakthrough was to realise that critically important is not the place where observations are made,[3] but what is observed and how it is observed (Bolduc, 2012). Early ethologists therefore demarcated their research area through focusing on the importance of natural causes and functions of behaviour, in contrast to the work of physiologists (interested in integrated physical movement) and behaviourists and psychologists (interested in the cognitive capabilities of organisms) (ibid).

Over time, ethology coalesced around a set of central questions posed by Tinbergen in the 1980s that focused on why and how different behaviours develop: these were gradually answered by academics who no longer saw themselves as tackling different aspects of the central questions, but instead tackling different questions (Griffiths, 2007). In addition, ethological research's defining characteristic – fieldwork – posed serious technical challenges: How could researchers separate and understand individual environmental influences (e.g. – predators' presence) on animal behaviour (Bolduc, 2012)? This led to research increasingly taking place in semi- or completely artificial situations. At the same time, related disciplines in other regions (such as North American experimental psychology) were moving into similar research areas and themselves growing in popularity. Subsequently, these meetings of disciplines have led to a blurring of distinctions between ethology and other disciplines and the rise of new disciplines like sociobiology, as well as the maturing of specific research areas like behavioural ecology into independent disciplines (Greenberg, 2012). Bolduc (2012) also notes that ethology appears in disciplines like cognitive ethology, applied ethology and neuroethology, which, while they stem from classical ethology, are independent disciplines.

What is clear is that the century-long characterisation of humans as exceptional animals that make use of interpretation, and animals as creatures only subject to evolutionary forces and the forces of nature and biology, which is the basis of the profound opposition between the social sciences and the natural sciences (Lestel, 2014), is now being challenged by both social and natural scientists. As Dowling et al. (2017) outline, there have been significant efforts in recent years from a diverse range of disciplines within the social sciences to change the relationships that researchers have developed with animals, both by using existing techniques in new ways or with a greater sensitivity to the more-than-human aspects of those techniques, and by developing new methods that address the performative and embodied elements of non-human lives.[4] We suggest that the main driver of this response is a greater awareness of interdisciplinary research, and because of physical and cultural shifts in society's engagement with the environment, resulting from recognitions of the Anthropocene. We now explore this in more detail through the example of recent studies of the honey bee, which exemplify efforts towards this broader approach. Our focus on honey bee decline illustrates why it is that when researchers start to engage with a specific nonhuman species, especially a species that is wild or semi-wild and therefore has entanglements with many other species and habitats, the resulting research can draw from multiple academic disciplines, as well as hitherto undervalued forms of knowledge about particular animal species, generated by humans.

The Need for the Two Species Ethnography Method: the Case of the Honey Bee

In 2006, a steep increase of honey bee colony deaths was identified in North America (vanEngelsdorp et al., 2006). When a parallel decrease in honey bee populations was identified in Europe, including the UK (Potts et al. 2010), the issue of insect pollinator decline became a major feature in the UK news (e.g. Carrington, 2014; Wells, 2007), and even featured in popular culture (e.g. the Bee Movie, Dreamworks Animation, 2007). The specific focus on the role of insects in food production (Grossman, 2013; Watanabe, 1994) reflected the rise of a "master frame" of food security in a time of human population growth, which has strongly influenced recent public policy (Mooney & Hunt in Ilbery, 2012).

The result of this policy and media interest was a surge in research into the causes of honey bee population losses, and into finding out just how dependent food production is on insect pollinators, such as honey bees. Both authors of this chapter were part of this "research moment" (Ward & Jones 1999, p. 301), carrying out research into honey bee (*Apis mellifera*) health and the beekeeper communities that care for them in the UK. Siobhan Maderson (SM) started from a social science approach, while Emily Adams (EA) took a natural science perspective. Yet our work gradually led us to a shared awareness that understanding honey bees, and other nonhuman species, necessitates engagement with a broad range of factors and sources of data. These include, but are not limited to, entomology, biology, agriculture, anthropology, history, Science and Technology Studies (STS), media studies and more. Both our studies were context-specific and motivated by the active environmental issue of pollinator declines and how government policy developed in response to these declines in the UK. The context we worked in was complex, rapidly evolving, and of public interest.

One of the most important outcomes of the "research moment" is that claims of honey bee declines, "global pollinator crises" and the end of food as we know it were hugely overblown. Although this was already recognised by some entomologists (e.g. Ghazoul, 2005), the complexity of the situation is still being explored by scientists and beekeepers. There are entire books to be written about the complexity of honey bee and other insect pollinator lives, but we briefly outline the main points here, within the context of TSE.

Central to TSE is a recognition of the multiple, varied humans interacting with, and affecting, a particular species. Therefore, we used interviews and ethnographic work with beekeepers, archival analysis of beekeepers' memoirs and beekeeping association histories, discourse analysis of media coverage and government policy documents, and literature reviews of natural science's research findings on bees and pollinators to understand the lived experience of our species of interest. From this, it became clear that the phrase "insect pollinators" is frequently used when talking only about the single species *Apis mellifera*. In order to be accurate when talking about insect pollinators, it is important to emphasise that honey bees are just one of thousands of species of insect pollinators, all of which contribute to wild-flower and crop pollination. For a variety of reasons, honey bees are the subject of a disproportionate amount of research and interest into their wellbeing, often at the cost of funding for other insect pollinators such as bumblebees, and the many species of solitary bees (with an attendant academic conflict over this bias, e.g. Corbet, 1991; Morse, 1991; Ollerton et al., 2012). The majority of research on honey bees tends to be in the natural sciences, reflecting the long fascination that humans have had with this social insect species.

Secondly, much of the current scientific research on honey bees (and other insect pollinators) engages with wider, often highly charged debates regarding food systems and agricultural practices in Europe and North America (Marshman et al., 2019; Shanahan, 2022). This situation may be familiar to other animal researchers, who can find their study species and its habitat forming part of a larger discussion with diverse stakeholders, within and beyond academia.

Finally, all the ecological and biological research into pollinator declines indicate that the causes are multiple, interlinked and scale-dependent, as well as vary across locations (Vanbergen and Insect Pollinators Initiative, 2013). For example, in the case of honey bee declines, it was found that in the USA, although initial efforts focused on relating the declines to specific causes such as viruses (e.g. Cox-Foster et al., 2007), the eventual conclusion is that Colony Collapse Disorder (CCD) is a result of an interaction between pathogens and other stress factors such as long-distance transport to pollinate different crops, a relatively modern activity that is the result of changing agricultural practices (vanEngelsdorp et al., 2010). In contrast, once researchers investigated the situation in the UK and Europe, bee declines since the 1980s were noted, coinciding with the arrival of a new parasitic pest of honey bees, *Varroa destructor*, which resulted in many hobby beekeepers ceasing their practice (Potts et al, 2010). Alongside this honey-bee specific research, and with a new perspective and interest in insect lives from the public and policy worlds, it was also noted that insect populations across all habitat types have dramatically reduced in number over the last 20–40 years in Europe, for causes that are not attributable to a single cause (Hallmann et al., 2017). This growing body of research into the causes of honey bee declines, and the complex interactions between them, has illustrated the shortcomings of conventional scientific inquiry as a method for understanding this species (Suryanarayanan & Kleinman, 2013; Maxim & Van der Sluijs, 2007, 2010). Importantly, it has also brought attention to significant human actors in the lives of these animals, including beekeepers, agrochemical retailers, policymakers, the media and environmental campaigners, to name but a few (Lezaun, 2011).

In order to understand what is happening to honey bees at a particular time and in a particular place (e.g. because you are a beekeeper, or because bees are declining, or you are interested in their wellbeing), you may need to investigate diverse topics such as:

- who actually manages the bees and what social dynamics flow through the beekeeping groups in your area (communities of practice theory)
- the historic trends and land change in areas where bees forage (historical geography)
- what has happened with the weather – both within a season, and with the climate across seasons (meteorology)
- what trade flows and management fashions there are that lead to bees and equipment being imported from overseas (economics)

- what public policy decisions have been made about pollinator health and land management (policy studies)
- On top of this, you will likely also need to become familiar with the life history and biology of the species itself, simply to understand what happens to a colony throughout a 12-month period (biology/zoology) and possibly also learn how to handle bees in order to get closer to them (beekeeping/ethology).

You may also need to engage with media studies, to investigate dissonance between public (mis)understandings and actual ecological realities (Smith, 2016; Wilson et al., 2017). As natural scientists struggle to devise field studies that replicate the lived environmental complexities of daily life for other species (as an example, see the debates around pesticides and their effects on pollinators documented by Maxim & van der Sluijs 2007, 2010), and as social scientists endeavour to transcend the limitations of human-centred methodologies (Dowling et al., 2017), being open to other forms of methods and data is critical. In the next section, we outline the TSE method which we offer towards this aim.

Two Species Ethnography in Practice

There is a tendency within the life and the social sciences to work with the Anthroparchal mindset, where nonhumans are institutionally interpreted and dominated through human attitudes and practices (Cudworth, 2014). However, nonhuman species are neither metaphors nor concepts, but living, real and embodied organisms (Pedersen & Stanescu, 2014). We believe that any study claiming to talk about human-nonhuman relations, while failing to directly engage with animals, is inevitably one-sided and imbalanced. TSE offers the chance to bring animals directly into a discussion which maintains a focus on the particular, singular species which a researcher is aiming to understand.

In our Two Species Ethnography practice, we specifically draw on existing, widely-used social science methods listed earlier to investigate the social-human aspects of honey bee lives, while also drawing on behavioural, ecological and biological work on bee health, nutrition, environmental and land management change to contextualise the human-honey bee relationship. The idea of working across academic disciplines dates back to the 1950s (Miller et al., 2008) but there is little consensus as to what "interdisciplinary" means in practice (Robinson, 2008), although this has not prevented a general increase in research which brings together a range of disciplines (Stirling, 2014; Mascia et al., 2003). The importance of engaging with social science is central to conservation, yet contrasting epistemologies are recognised as a challenge (Moon & Blackman 2014). MacMynowski (2007) suggests that there are two key ways of talking about interdisciplinarity: as a tool or method to answer specific questions, and as a research aim in its own right, described by Robinson (2008) as "issue-driven" and "discipline-based" interdisciplinarity, respectively.

The way TSE seeks to reach between academic disciplines therefore builds on this tradition, but the focus on the nonhuman species of interest as well as its environmental context and its human cohabitors puts an explicit focus on understanding things from the perspective of the individual animal/colony: a perspective often lacking in interdisciplinary research. Thus, TSE is pushing at the boundaries of what is regarded as "interdisciplinarity" (itself a coalescing field around environmental topics) to draw in HAS to these discussions. By keeping a focus on a specific case study (e.g. a species under threat of decline), TSE helps scholars navigate the conceptual breadth of existing HAS approaches like MSE while also engaging with diverse other disciplines as the case study context requires.

In our TSE practice, we resonate with the desire of Critical Animal Studies researchers to radically change human-animal relations (Taylor & Twine, 2014), as our interest in the honey bee as a research topic came about because of the concerns around population declines. Our research took place in a time of loss and fear – as does much ecological, conservation and environmental research. In the environmental and biological sciences, there is often a desire to study species that are suffering/being affected by some aspect of anthropogenic environmental change, which resonates clearly with the roots of HAS in the ethics of domestic animal care.

Having said this, why do we bother to engage with the human? Why not simply study and understand the honey bee's complex ecological and environmental entanglements, which seem to include all-encompassing explanations for their loss?[5] As we already discussed, the context of our study, the UK, is one where human action is closely woven through the environment, and where the majority of honey bees are in some way managed by people. This is very common for many nonhuman species, especially in Europe (Taylor & Hamilton, 2014): they are in some way owned, managed, or live on land owned and managed by people. Even those bees which are not being actively managed by beekeepers are being affected by anthropogenic factors. Thus, humans often serve as both intermediaries for observing and documenting other species' existence, and as active agents impacting other species through relationships between humans and other animals. Beekeepers, farmers, gardeners and fishers are gatekeepers to animal species. Ethnographic engagement with this second species – namely, humans – is central to understanding the species in question, and is the hallmark of TSE.

For example, as we found in our case study of honey bees, the disproportionate focus on honey bees over the thousands of other pollinators that have also been declining (Smith & Saunders, 2016) reflects both humans' social engagement with this species, and the fact that there is a group of people who live and work closely with honey bees, namely beekeepers. This group is attuned to honey bees, and attentive to their daily and annual behaviour and lives (Adams, 2016), as observers of honey bees (learning about them), as active agents in shaping honey bee lives for human purposes (for honey and wax production, and pollination), and, most recently, as stewards of honey bee health. They are therefore well positioned to identify varied, relevant issues impacting bees, and are able to act as intermediaries between human society (its governance, media, scientists and so on) and honey bees. Beekeepers' entanglement with bees affects both species (bee and human). It is therefore clear that any HAS researcher interested in honey bees will engage with beekeepers in some way. We propose that an ethnographic approach best reflects the rich history and diversity of forms of engagement with honey bees (Figure 5.1).

We chose the term "ethnography" to describe our TSE practice because ethnography is "the art and science of describing a group or culture" (Fetterman, 2004, p. 328) to provide "a detailed, in-depth description of

Figure 5.1 Photo by Emily Adams. Successful beekeeping (defined as colonies surviving through winter, producing honey, pollinating crops, etc.) requires beekeepers to regard the wellbeing of the colony from inside and out.
Monitoring internal wellbeing involves the close inspection of a colony of
honey bees, requiring the beekeeper to be patient, to look closely, and to
seek an awareness of the bees and their condition through paying attention
to their behaviour and patterns within the hive such as stores of food, presence or absence of eggs, presence of the queen among other factors. This
internal condition is affected by external factors such as availability of
flowers (sources of nectar and pollen), weather, and seasonal patterns of
colony development. Thus, beekeepers do more than just count bees, they
have to both pay attention to but also transcend the literal living organisms
to also consider the situation of the colony through time and in place. This
process is a challenge to the strong focus in the media and in policy on the
usefulness of honey bees to humans, and refocuses attention on the needs
of the honey bees themselves - a transcendence of the human-bee divide.

everyday life and practice" (Hoey, 2014). The traditional social sciences have often silenced nonhuman and non-literate voices through its emphasis on numerical and textual methods (Dowling et al., 2017). However, the original purpose of ethnography is to study and write about people in their natural setting through direct interaction with them (Taylor & Hamilton, 2014). In the case of HAS, the group or culture being described is that of a different species – reflecting the growing recognition in the social sciences that nonhuman animals are social beings in their own right (Taylor & Twine, 2014). With TSE, the focus is on two species: humans and a nonhuman of notable interest to some humans.

Ethnography with human groups commonly draws on a mix of techniques to help reveal patterns about that community (Herbert, 2000).[6] Two Species Ethnography specifically draws on techniques and literature from both natural and social science research traditions to enable us to both engage with animal species, whether individuals of a species (e.g. large mammals) or groups (e.g. bee colonies), while recognising and engaging with human behaviours and knowledge that impact and illuminate the species in question. It also acknowledges, and actively engages, with multiple forms of human expertise and understanding of an animal species in question. TSE offers practical methodological boundaries for studying a specific species, while also providing an egalitarian, multidisciplinary approach to engaging with multiple human perspectives and experiences of a particular species. We recognise the challenges of prioritising human voices when understanding any animal. However, TSE is particularly relevant when considering those species who may be of great interest to some humans yet are often outside the common bias towards companion or other very charismatic species.

Within the term "Two Species Ethnography" we blend methods such as interviews with key participants, participant observation (of both the human and the nonhuman) and archival research with more traditional natural science methods of observing and studying nonhumans, such as capturing them and studying their ecological engagements. Specifically, and importantly, engaging with key human participants allows triangulation of data from the observations and experiences of researchers, or from multiple text sources, enabling researchers to go below the surface and into the "mess" of the field (Taylor & Hamilton, 2014). Interviews with people who have spent years intimately observing and working with other species can provide valuable understanding of that species' needs, and behavioural responses to external conditions. We also note the relevance of natural science methods in understanding behaviours and information relevant to bees – our particular species of interest. As a concrete example from our research practice of how this methodological mixing happens, EA was interested in honey bee nutrition and how it was affected by agriculture. EA captured pollen from honey bees, inspecting these under a microscope (following techniques used in palaeontology) while also talking to beekeepers about their perceptions of what plant species their bees were likely to visit and how this changed over time (interviews), observing beekeeper practices such as supplementary feeding with

sugar syrup (participant observation) as well as doing direct field observations of available plant species (botany). This allowed her to combine
human-observed temporal change in land use with the lived experience of
honey bees.

Critics of the current status quo in HAS argue that there is still an emphasis on engaging with the human interpretation of animals (Hodgetts &
Lorimer, 2015). We would argue that while this is something to be aware of
as a researcher, many key human participants in a Two Species Ethnography,
such as the long-term beekeepers we interviewed and worked with as part of
our research, are quick to emphasise the importance of empathetic observation of the nonhuman participants (e.g. honey bees) in their environment as
a key part of engaging with that species – something also familiar to natural
scientists who observe their study species in-situ or ex-situ. Altering one's
perception to attune human environmental interpretation with that of other
species is explored by veterinarian, barrister and polymath Foster (2016),
who addresses key differences between species in their neurological processing as a factor that can, ultimately, never be overcome – although any human's
attempt to explore another species' lived experience is profoundly enriching,
and humbling. While there is a human tendency to assume that the world we
see is the way the world is, we know from experiments and observation by
beekeepers and scientists that, for example, honey bees, unlike humans, can
see on the UV spectrum, and cannot see various shades of red. We inhabit
two different yet parallel worlds of colour and pattern when it comes to
admiring colourful flowers, highlighting that vision is a property of the
viewer and the object being viewed, but also of the relationship between them
(Raffles, 2010). When interspecies differences such as body size, eye structure,
mobility, and social structure are taken into consideration, the distance
grows. With this distance, there is the realisation that humans and bees (and
by extension all nonhuman species) exist in parallel, invisible yet intersecting
worlds. TSE, by drawing together the detailed observations of nonhuman
animals from natural scientists with the in-depth research and observations
by other human stakeholders, helps to translate these differences in worldviews and experience, giving weight to the complex, and at times unknowable,
experiences of the nonhumans.

Observation is the key to attempting to bridge this gap. Most of what is
known about honey bees, for example, comes from decades of meticulous
biological observation and manipulation by interested amateurs, early natural historians and more recently by professional scientists, dating back to the
earliest known observations about honey bee lives by Aristotle in Ancient
Greece (Crane, 2011). Without experiments and observation by scientists and
beekeepers (who may inhabit both roles at once), honey bees would still be
thought to be led by king bees (disproven in the 1600s), that honey and wax
emerge miraculously from the environment (the complex processes of wax
and honey production being finally clarified in the 1700s). Such misunder-
standings continue, with the mysterious deaths of colonies in the mid-2000s
considered by some to be the result of alien abductions rather than incredibly

complex, interlinked ecological and management factors combined with unexpected effects of pesticides (Vanbergen and Insect Pollinators Initiative, 2013). While this knowledge may not tell us what it is like to live in a colony, it does allow us to ensure our imaginings and efforts to experience that world are more accurate. Many beekeepers during interviews emphasised the central importance of their physical and emotional engagement with bees in understanding bee health and needs. Examining a frame of bees when looking for the queen requires calm, still concentration, which can be considered intra-species mindfulness (Moore & Kosut, 2013). A relationship is generated which affects all participants. Such reflection serves to heighten human sensitivities to the lived experience of other species.

For long-term beekeepers, the experience of observing and engaging with this species provides unique insights into its behaviour, and its response to a plethora of environmental externalities – many of which are anthropogenic. Interviews and beekeeping associations' archives note that beekeepers are radically changing the timing of their annual practices, in response to changes in their bees' breeding cycle, which are in turn caused by significant phenological changes (Maderson & Wynne-Jones, 2016; Adams, 2016; Phillips, 2014). Understanding these changes as a researcher interested in why a change is happening requires biological and ecological knowledge, as well as engaging with long-term human observers of animals. TSE therefore enables us to temporally contextualise animal behaviours and human behaviours in a changing world.

Ethnographic techniques such as archival research and participant observation are rich sources of data but must also be carried out critically, with concerns around bias, sampling, timing of observations and changes in what is considered "ethnography" being occasionally articulated from within the social science disciplines (Hammersley, 2006). Specifically, the ability to record material for detailed analysis allows researchers to lose the holistic perspective on a situation (Hammersley, 2006) – a risk that is especially possible in a context such as observing and understanding animals where environmental factors may play a key role. Due to TSE's gathering diverse data sources via multiple methods, the risks of bias are less than may be found when relying on a single method for understanding a species. For example, interviewees mentioned changes in the weather and how that affected their bees. These comments could be compared with archives and texts which supply phenological data and are also linked to farming and land management information which might indicate changes in land use over time in response to economic and climatic changes.

An important aside to this point is that although TSE is a flexible methodical approach, one key point - often overlooked in methods guidance - is the importance of timing. We see this as key in guiding ecological, ethological and social scientific methods, because all of us, human and nonhuman, are shaped by the seasonal pattern of life on Earth. When is the right time to collect certain types of data? Species have chronological patterns that must be addressed in TSE. Honey bees follow an annual cycle of colony growth and contraction, while beekeepers also follow an annual cycle of availability for interviews, participant observation and other data collection methods. An effective piece of research into human-animal relations will have to take account of the seasonal variations and patterns of behaviour - in part because they shape the life of animals and people, but also from a strictly practical perspective. Cultivating and applying such a sensitivity to research is but one step in accurately reflecting the lived reality of our nonhuman colleagues and companions. Their world is our world, and to truly understand them and our relationship with them, it is necessary to bring in, and work from, a diversity of viewpoints.

Conclusion

Two Species Ethnography is an approach that equips us with multiple proxy tools that, when used together, can build an emotional, ethical and political connection with species that may be very difficult for us to empathise with or understand. TSE prioritises the lived experience of species in a diverse and anthropogenically changing global environment as a starting point which drives the choices of varied methods.

While the social sciences should be congratulated for the "animal turn", it will enhance research and understanding to remember that the natural sciences have been working on, and with, animals for centuries. HAS has an important role to play in a wider research agenda but needs other methods to broaden its appeal and relevance. As we (as a global society) become conscious of our global impact, using TSE can advance the development and operationality of Human-Animal Studies to a broad audience, including ethicists, linguists, conservationists and land managers. Current HAS seems to be primarily focused on human relationships with other mammals. By broadening its methods and resultant approach, HAS can ultimately inspire and guide many potential students and researchers in a wider engagement with the environmental context of human and nonhuman lives. TSE offers one route towards a multidisciplinary methodology that explicitly considers the entanglement of humans, nonhumans and their environment.

By drawing on methods from both the social sciences' long engagement with human and society questions, and the natural sciences' long engagement with understanding the fundamentals of nonhumans, researchers in the HAS field are able to create a well-rounded, more accurate understanding of nonhumans, and human influences on and relationships with them (Lestel et al., 2006). In the case of honey bees and insect pollinator decline, understanding and ultimately reversing current insect pollinator decline requires us to engage with agricultural practices, climate patterns and global trade. It also necessitates investigating environmental values, and behaviours of beekeepers, consumers and land managers.

The close relationship between animals and humans, and particularly understanding animals (or even plants, microbes and other denizens of the natural world) goes beyond talking about their role in our lives. Humans have consistently influenced the lives of the nonhumans around us – sometimes purposefully, through selective breeding, and other times inadvertently, such as by releasing hormone-disrupting chemicals, or destroying/altering habitats. Humans who engage deeply with animals, whether practitioner, natural or social scientist, can testify to the transformative quality of this engagement on human values, identity and understanding of both the observer and the observed. While this phenomenon is gaining prominence in recent Western science literature, anthropologists and writers on Traditional Environmental Knowledge can attest to humans' long history of blurred distinctions between humans and other species (Nadasdy, 2007).

TSE is a broad, mixed-methods approach which is highly relevant, practical and applicable for studying a wide range of animals. It also has the potential capacity to bring HAS to a wider readership and stimulate deeper understandings of a multitude of species. While mammals and other charismatic species dominate the research agendas and wider writings, we propose that TSE facilitates a more accurate, deeper understanding of species which receive far less of our attention.

What we propose here as TSE, and the context in which we present it (that of honey bees and beekeepers) is only one example where it could be useful. While not every HAS researcher will be working on a particular species of interest to them, for those students and researchers who are focusing on a single species, TSE can be a useful method to understand the complex relationships between humans and the species in question, by addressing the diverse sources of understanding of another species. If researchers are interested in the Monarch butterfly, for example, then it is important to engage with the reality of monarch butterfly conservation programmes being affected by popular media and politically expedient, photogenic actions (Gustafsson, 2017). A researcher studying orangutans' behaviour could benefit from engaging with Palm Oil plantation expansion rates, and campaign groups that encourage supermarkets to use substitutes to palm oil. Importantly, TSE also provides a method for understanding temporal elements of species' changes, which are at risk of being overlooked via singular methods. Such temporal factors are highly relevant as we strive to understand how climate change and other anthropogenic environmental changes are impacting diverse species.

Notes

1 A linguistic point of discussion is the word 'honey bee' or, as it is also commonly spelt, 'honeybee'. Entomological nomenclature says that if the common name of an insect implies what it is then the words are separate, and if not then the words are combined (Dennis, 2011). Thus, a honey bee is a true bee which creates honey and so the words are separate, whereas a butterfly is not a fly and so the words are combined (Dennis, 2011). However, in North America, regardless of this, honeybee is typically written as one word, whereas in Europe the opposite is true (Dennis, 2011). In this chapter the European tradition is followed.

2 See Lorimer, 2007 for a discussion about nonhuman charisma.

3 It is important to note here that while late 20th century ethologists moved away from focusing on where animal observation was done onto what was observed, HAS scholars and ethicists have returned to the question of research location as central to understanding human-animal relations (Birke, 2014). In particular, Critical Animal Studies, and its connections to the animal rights movement, make the link between animal exploitation/domination by humans, and researcher responsibility, animal agency and ethics of consent in research more visible (Birke, 2014). This raises important questions about the limitations of research on animals, and especially the focus of ethologists on breaking down animal behaviour into separate parts, rather than using this sort of observation to understand the dynamic processes that make up a human-animal relationship or which can indicate animal preferences, views and opinions (Birke, 2014; Dawkins, 2006).

4 This Anglophone work resonates with, but has widely ignored, a continental European group of authors who have been working on bridging the ethnograph-ic-ethological divide for several decades in a quest to reimagine research on and engagement with (both personal and academic) human-animal relations, a group that Bussolini (2013) helpfully describes.

5 In the context of honey bees, the natural sciences have been relatively slow to take a 'human turn': the anthropogenic influence on honey bee health and wellbeing has only recently become an important aspect of scientific work on honey bees. The recently coined phrase 'Darwinian beekeeping' (Seeley, 2017; Neumann and Blacquiere, 2016) indicates that honey bee scientists are only now turning to look at the effects of hundreds of years of honey bee breeding and management by beekeepers on bees' evolution and adaptation to anthropogenic environments, and their health and wellbeing. Similarly, there are also social movements within UK beekeeping that are actively rejecting decades of recommended methods of beekeeping practice in favour of methods that they regard as more reflective of the bee in its 'natural' (i.e. non-managed) environment, and which emphasise honey bee welfare over honey production and yield. These shifts within beekeep-ing reflect changing perceptions and a stronger identification with the bees and their interests/desires (as far as these can be understood) – a gradual shift towards the starting point of HAS as a discipline.

6 Ethnography has been used to study fish (Bear & Eden, 2011), honey bees (Phillips, 2014), and corncrakes (Lorimer, 2008).

References

Adams, E. (2016) How to become a beekeeper: Learning and skill in managing hon-eybees. *Cultural Geographies*, 25(1): 31–47.

Bear, C., Eden, S. (2011) Thinking like a fish? Engaging with nonhuman difference through recreational angling. *Environment and Planning D: Society and Space*, 29(2): 336–352.

Birke, L. (2014) Listening to voices: On the pleasures and problems of studying human-animal relationships. In: Taylor and Twine (eds), *The Rise of Critical Animal Studies*. Routledge, U.K.

Bolduc, J.-S. (2012) Behavioural ecology's ethological roots. *Studies in History and Philosophy of Biological and Biomedical Sciences*, 43(3): 674–683.

Buller, H. (2014) Animal geographies I. *Progress in Human Geography*, 38(2): 308–318.

Bussolini, J. (2013) Recent French, Belgian and Italian work in the cognitive science of animals: Dominique Lestel, Vinciane Despret, Roberto Marchesini and Giorgio Celli. *Social Science Information*, 52(2): 187–209.

Carrington, D. (2014) UK faces food security catastrophe as honeybee numbers fall, scientists warn. *The Guardian*, 8th January 2014. http://www.theguardian.com/environment/2014/jan/08/uk-food-security-honeybees (accessed 10/9/2014).

Corbet, S. (1991) Applied pollination ecology. *Trends in Ecology and Evolution*, 6: 3–4.

Cox-Foster, D., Conlan, S., Holmes, E., Palacios, G., Evans, J., Moran, N., et al. (2007) A metagenomic survey of microbes in honey bee colony collapse disorder. *Science*, 318: 283–287.

Crane, E. (2011) *The World History of Beekeeping and Honey Hunting*. International Bee Research Association, Cardiff.

Crist, E. (2004) Can an insect speak? The case of the honey bee dance language. *Social Studies of Science*, 34: 7–43.

Cudworth, E. (2014) Beyond speciesism: Intersectionality, critical sociology and the human domination of other animals. In: Taylor and Twine (eds), *The Rise of Critical Animal Studies: From the margins to the centre*. Routledge, UK, London, pp. 19–35.

Cudworth, E. (2016) A sociology for other animals: Analysis, advocacy, intervention. *International Journal of Sociology and Social Policy*, 36(3/4): 242–257.

Dawkins, M. (2006) Through animal eyes: What sentience can tell us. *Applied Animal Behaviour Science*, 100: 4–10.

Dennis, B. (2011) Ask the Experts: To bee or not too bee—Honeybee, honey bee or honey-bee? *BBKA News*, 118: 25.

Dowling, R., Lloyd, K., Suchet-Pearson, S. (2017) Qualitative methods II: 'More-than-human' methodologies and/in praxis. *Progress in Human Geography*, 41(6): 823–831.

Dreamworks Animation. (2007) *The Bee Movie* [film]. Dreamworks Animation, California.

Fetterman, D. (2004) Ethnography. In: Lewis-Black, M., Bryman, A., Liao, T. (eds), *Encyclopedia of Social Research Methods*. Sage Publications Inc., London. pp. 328–333.

Foster, C. (2016) *Being a Beast*. Profile Books, U.K.

Ghazoul, J. (2005) Buzziness as usual? Questioning the global pollination crisis. *Trends in Ecology and Evolution*, 20: 367–373.

Greenberg, G. (2012) *Comparative psychology and ethology*. In: Seel, N. M. (eds), *Encyclopedia of the Sciences of Learning*. Springer, Boston, MA.

Griffiths, P. (2007) History of ethology comes of age. *Biological Philosophy*, 23: 129–134.

Grossman, E. (2013) Declining bee populations pose a threat to global agriculture. *Yale Environment 360*, 30th April. http://e360.yale.edu/feature/declining_bee_populations_pose_a_threat_to_global_agriculture/2645/ (accessed 10/9/2014).

Gustafsson, K. M. (2017) Narrating the monarch butterfly: Managing knowledge complexity and uncertainty in coproduction of a collective narrative and public discourse. *Science Communication*, 39: 492–519.

Hallmann, C. A., Sorg, M., Jongejans, E., Siepel, H., Hofland, N., Schwan, H., et al. (2017) More than 75 percent decline over 27 years in total flying insect biomass in protected areas. *PLoS ONE*, 12(10): e0185809. https://doi.org/10.1371/journal.pone.0185809

Hammersley, M. (2006) Ethnography: Problems and prospects. *Ethnography and Education*, 1(1): 3–14.

Herbert, S. (2000) For ethnography. *Progress in Human Geography*, 24: 550–568.

Hill, R., Adem, Ç., Alangui, W. V., Molnár, Z., Aumeeruddy-Thomas, Y., Bridgewater, P., et al. (2020) Working with indigenous, local and scientific knowledge in assessments of nature and nature's linkages with people. *Current Opinion in Environmental Sustainability*, 43: 8–20.

Hodgetts, T., Lorimer, J. (2015) Methodologies for animals' geographies: Cultures, communication and genomics. *Cultural Geographies*, 22(2): 285–295.

Hoey, B. (2014) A simple introduction to the practice of ethnography and guide to ethnographic field notes. *Marshall University Digital Scholar*, 1:10.

IPBES. (2019) Summary for policymakers of the global assessment report on biodiversity and ecosystem services of the Intergovernmental Science-Policy Platform on Biodiversity and Ecosystem Services. In: S. Díaz, J. Settele, E. S. Brondízio, H. T. Ngo, M. Guèze, J. Agard, A. Arneth, P. Balvanera, K. A. Brauman, S. H. M. Butchart, K. M. A. Chan, L. A. Garibaldi, K. Ichii, J. Liu, S. M. Subramanian, G. F. Midgley, P. Miloslavich, Z. Molnár, D. Obura, A. Pfaff, S. Polasky, A. Purvis, J. Razzaque, B. Reyers, R. Roy Chowdhury, Y. J. Shin, I. J. Visseren-Hamakers, K. J. Willis, and C. N. Zayas (eds), *IPBES secretariat*. Bonn, Germany.

Ilbery, B. (2012) Interrogating food security and infectious animal and plant diseases: A critical introduction. *The Geographical Journal*, 178: 308–312.

Jensen, P. ed., (2017) *The ethology of domestic animals: An introductory text*. CABI, Wallingford.

Lave, R. (2014) Engaging within the academy: A call for critical physical geography. *ACME: An International E-Journal for Critical Geographies*, 13(4): 508–515.

Lestel, D., Brunois, F., Gaunet, F. (2006) Etho-ethnology and ethno-ethology. *Social Science Information*, 45(2):155–177.

Lestel. D. (2014) Toward an ethnography of animal worlds. *Angelaki: Journal of the Theoretical Humanities*, 19(3): 75–89.

Lezaun, J. (2011) Bees, beekeepers and bureaucrats: Parasitism and the politics of transgenic life. *Environment and Planning D: Society and Space*, 29: 738–756.

Lockwood, J. (1987) The moral standing of insects and the ethics of extinction. *Florida Entomologist*, 70(1): 70–89.

Lorimer, J. (2007) Nonhuman charisma. *Environment and Planning D: Society and Space*, 25(5): 911–932.

Lorimer, J. (2008) Counting corncrakes: The affective science of the UK corncrake census. *Social Studies of Science*, 38(3): 377–405.

MacMynowski, D. (2007) Pausing at the brink of interdisciplinarity: Power and knowledge at the meeting of social and biophysical science. *Ecology and Society*, 12: 20.

Maderson, S., Wynne-Jones, S. (2016) Beekeepers' knowledges and participation in pollinator conservation policy. *Journal of Rural Studies*, 45: 88–98.

Mascia, M., Brosius, J., Dobson, T., Bruce, C. Forbes, B., Horowitz, L., et al. (2003) Conservation and the social sciences. *Conservation Biology*, 17: 649–650.

Maxim, L., van der Sluijs, J. (2007) Uncertainty: Cause or effect of stakeholders' debates? Analysis of a case study: The risk for honey bees of the insecticide Gaucho®. *Science of the Total Environment*, 376: 1–17.

Maxim, L., van der Sluijs, J. (2010) Expert explanations of honeybee losses in areas of extensive agriculture in France: Gaucho® compared with other supposed causal factors. *Environmental Research Letters*, 5: 014006.

Marshman, J., Blay-Palmer, A., Landman, K. (2019) Anthropocene crisis: Climate change, pollinators, and food security. *Environments*, 6(2): 22.

Millennium Ecosystem Assessment. (2005) *Ecosystems and human well-being: Biodiversity Synthesis*. World Resources Institute, Washington D.C.

Miller, T., Baird, T., Littlefield, C., Kofinas, C., Chapin Iii, F., Redman, C. (2008) Epistemological pluralism: Reorganising interdisciplinary research. *Ecology and Society*, 13: 46.

Moon, K., Blackman, D. (2014) A guide to understanding social science research for natural scientists. *Conservation Biology*, 28(5): 1167–1177.

Moore, L.J., Kosut, M. (2013) *Buzz: Urban Beekeeping and the Power of the Bee*. New York University Press, New York.

Morse, R. (1991) Honeybees forever. *Trends in Ecology and Evolution*, 6: 337–338.

Nadasdy, P. (2007) The gift in the animal: The ontology of hunting and human–animal sociality. *American Ethnologist*, 34: 25–43.

Ollerton, J., Price, V., Armbruster, W., Memmott, J., Watts, S., Waser, N., et al. (2012) Overplaying the role of honey bees as pollinators: A comment on Aebi and Neumann (2011). *Trends in Ecology and Evolution*, 27: 141–142.

Pedersen, H., Stanescu, V. (2014) Conclusion: Future directions for critical animal studies. In: Taylor and Twine (eds), *The Rise of Critical Animal Studies*. Routledge, U.K.

Phillips, C. (2014) Following beekeeping: More-than-human practice in Agrifood. *Journal of Rural Studies*, 36: 149–159.

Raffles, H. (2010) *Insectopedia*. Pantheon Books, New York.

Robinson, J. (2008) Being undisciplined: Transgressions and intersections in academia and beyond. *Futures*, 40: 70–86.

Shanahan, M. (2022) Honey bees and industrial agriculture: What researchers are missing, and why it's a problem. *Journal of Insect Science*, 22(1):14.

Smith, T. J., Saunders, M. E. (2016) Honey bees: The queens of mass media, despite minority rule among insect pollinators. *Insect Conservation and Diversity*, 9(5): 384–390.

Stirling, A. (2014) Disciplinary dilemma: Working across research silos is harder than it looks. *The Guardian*, 11th June 2014. http://www.theguardian.com/science/political-science/2014/jun/11/science-policy-research-silos-interdisciplinarity (accessed 25/9/2014)

Suryanarayanan, S., Kleinman, D. (2013) Be(e)coming experts: The controversy over insecticides in the honey bee colony collapse disorder. *Social Studies of Science*, 43: 215–240.

Taylor, N., Twine, R. (2014) *The Rise of Critical Animal Studies: From the margins to the centre*. Routledge, U.K.

Taylor, N., Hamilton, L. (2014) Investigating the other: Considerations on multi-species research. In Martin Hand and Sam Hillyard (eds), *Big Data? Qualitative Approaches to Digital Research*. Emerald Group Publishing Limited, pp. 251–271.

Vanbergen, A., Insect Pollinators Initiative. (2013) Threats to an ecosystem service: Pressures on pollinators. *Frontiers in Ecology and the Environment*, 11: 251–259.

VanEngelsdorp, D., Cox Foster, D., Frazier, M., Ostiguy, N., Hayes, J. (2006) 'Fall-Dwindle Disease': Investigations into the causes of sudden and alarming colony losses experienced by beekeepers in the fall of 2006. Preliminary Report: First Revision. Florida Department of Agriculture, Florida.

VanEngelsdorp, D., Hayes, J., Underwood, R., Pettis, J. (2010) A survey of honey bee colony losses in the United States, fall 2008 to spring 2009. *Journal of Apicultural Research*, 49: 7–14.

Ward, K., Jones, M. (1999) Researching local elites: Reflexivity, 'situatedness' and political-temporal contingency. *Geoforum*, 30: 301–312.

Watanabe, M. (1994) Pollination worries rise as honey bees decline. *Science*, 265: 1170.

Wells, M. (2007) Vanishing bees threaten US crop. *BBC News*, 11th March 2007. http://news.bbc.co.uk/2/hi/americas/6438373.stm (accessed 10/9/2014).

Wilson, J. S., Forister, M. L., Carril, O. M. (2017) Interest exceeds understanding in public support of bee conservation. *Frontiers in Ecology and the Environment*, 15(8): 460–466.

6 Trekking a Predator's Journey

Paths through the Greater Yellowstone Ecosystem

Hannah Jaicks-Ollenburger

Late in the summer of 2013, I was sitting in a parking lot by the Jackson Hole community centre and waiting for my carpool to pick me up and take me to the Gros Ventre range. Just south of Grand Teton National Park, the Gros Ventre range is an area where several environmental partners were hosting a public hike to spread the word about their involvement with a large-scale wildlife campaign, TrekWest. As I waited and picked at my old boots that were duct-taped together, John Davis walked up to wait beside me. John had just reached Jackson Hole after walking, biking and paddling his way up the chain of mountains that constitute the Spine of the Continent. Also known as the "Western Wildway", the Spine is a North American wildlife corridor that stretches from the Sonoran Desert of Mexico through the Northern Rockies of Canada (Hannibal, 2012). This wildlife corridor boasts the Greater Yellowstone Ecosystem (GYE) near its crown, and it is an essential habitat for humans and wide-ranging predators alike. John had been undertaking TrekWest as a trailblazing approach to raise awareness about predator conservation.

At the time, I was still attempting to reconcile recent observations from my first experience of a hostile public meeting on the 2013 wolf harvest quota, with my growing understanding about the inadequacy of the current decision-making processes in the GYE on predator management. I was considering this matter as we bounced along the unpaved road to the Gros Ventre range, but my attention was distracted by the man sitting next to me, who I finally realised was John. After being introduced, we talked about the mission and concept of his journey as we hiked along the Path of the Pronghorn, a route through which antelope and other nonhuman animals traverse when they migrate north from the Upper Green River Valley of Wyoming each spring. We spent the day hiking around red-faced cliffs and discussing our backgrounds as we travelled the same paths that wildlife have travelled for more than 6,000 years (Figure 6.1).

Like many portions of the Western Wildway, this area and its carnivores are at risk of extinction due to the social and environmental hazards of real-estate development, increasing human populations and imbalanced predator–prey relationships (White, Garrott & Plumb, 2013). This geography has resulted in divisive social conflicts about these nonhuman animals' management, as well as physical conflicts when they come into contact with

DOI: 10.4324/9781351018623-8

Figure 6.1 Photo by Hannah Jaicks-Ollenburger. Sign at the trailhead of the Path of the
Pronghorn. The adjacent sign was destroyed by bullets the previous week.

humans over the course of daily life. These conflicts are physical and symbolic expressions of larger concerns that remain unaddressed by policy measures, and which polarise public debates regarding wildlife management
amidst a changing environment. During our trek, I began to realise that
research was needed to reveal the limitations of current scientific and policy
paradigms and develop new avenues for incorporating the stakes of nonhuman animals in debates about the land and its inhabitants. So, by the end of
the hike, an idea took hold. I arranged to join John and Ed George, the filmmaker accompanying him, for the GYE portion of TrekWest. This involvement was my opportunity to traverse a predator's journey and experience
what it takes to navigate these hazards on the landscape, from the perspective
of a grizzly, wolf or cougar.

There was a two-pronged challenge I would come to understand in undertaking a trek of this nature. There was the physical component of surviving
the distances and environmental challenges that these nonhuman animals

cover on their migratory routes, though at a much smaller scale than for a wolf or other predator. There was also the social challenge of translating the significance of that physical feat to humans from different perspectives in communities along the route. My trek through the GYE landscape, in which I encountered antagonistic community members, reckless tourists, unpredictable weather and treacherous road crossings, was a challenging one. Yet, these challenges only served to underscore my burgeoning awareness of the obstacles, material and symbolic, a nonhuman animal encounters on its paths through the GYE.

This chapter traces these obstacles and their implications for predators' survival on the landscape through an interwoven discussion of theory and practice. Specifically, I begin by mapping my methodological approach, which must be understood as a basis for why I attended to the nonhuman animal "Other" when addressing human-wildlife conflicts. This framing is followed by a theoretical discussion of feminist science studies and animal geography, two disciplines that fall under the much broader field of human-animal studies. The vignettes that segment each section illustrate the experiential quality of my study, as this chapter is intended to achieve new means for examining and addressing interspecies conflicts as well as broaden the field of human-animal studies. Through this chapter, I reveal how and why humans' conflicts with nonhuman predators are so persistent and prevalent in order to identify the transformations in science and policy necessary for mediating coexistence amongst *all* stakeholders under contemporary conditions of environmental stress.

Methodological Approach

Methods of research on predator conflicts, and, more broadly, human–nonhuman animal relations have historically operated from a human-centred standpoint with limited recognition of nonhuman animals and the interconnectedness of nonhuman creatures to our lives and identities. To overcome concerns of marginalising the needs of predators, this study that I conducted for my doctoral research employed ethnographic methods to understand humans' contested relationships with one another and the predators of the GYE. To contextualise the theoretical discussion and the first-person narratives I share in this chapter, I first offer a short overview of my research site, the human and nonhuman animal actors in the area, and the approach that I took to carrying out my work. This overview is intended to provide the reader with a foundation for understanding my later discussion of the theoretical framings of feminist science studies and critical animal geography and their collective implications for addressing these contested relations.

The Place

Humans are expanding their control of the environment at a pace and scale that continues to degrade habitats and perpetuate conflicts across species (Rasker & Hansen, 2000). These conflicts are pervasive in the American

West, where disputes over resources are fiercely contested as climate change further fragments the landscape. At the heart of these disputes is the Greater Yellowstone Ecosystem (GYE), a 40,000 m² region considered hallowed ground for the global conservation community (Berger, 1991, 2009; US National Park Service, 2019). The terrain of the ecosystem spans nearly 20 million acres that encompasses private lands, Yellowstone and Grand Teton National Parks, portions of six national forests and five national wildlife refuges throughout Wyoming, Idaho and Montana (Figure 6.2).

The Predators

Further defining the GYE is its inimitable capacity to support large carnivores (particularly grizzly bears, wolves and cougars) that struggle to migrate, live and reproduce amidst an area undergoing rapid urbanisation

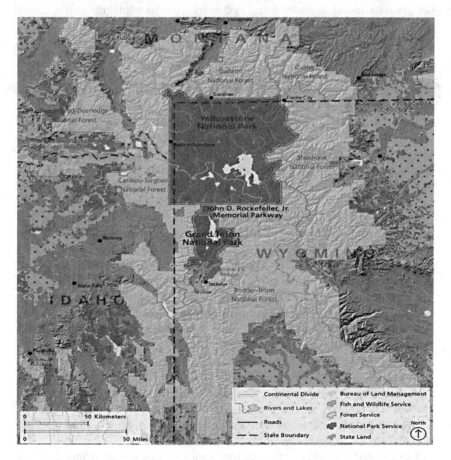

Figure 6.2 Study site of the Greater Yellowstone Ecosystem including Montana, Idaho, and Wyoming. Map and data from the U.S. National Park Service (2019).

(Blanchard & Knight, 1991). This geography has produced numerous conflicts across species because there is not sufficient space or quality resources to accommodate everyone's needs (Heinen, 2007; Jobes, 1991, 1993). The GYE is therefore a critical site for understanding the implications of human development in relation to large carnivore conservation within the broader context of climate change.

As a response to this imperative, I systematically examined humans' conflicts with the GYE's large carnivores. Grizzly bears, wolves and cougars are wide-ranging predators who require vast areas of land and resources to survive, and their presence is indicative of an ecosystem's stability. As keystone species, they exert a trophic cascade, a mechanism involving top-down predation that structures the ecological balance of the GYE (Smith & Ferguson, 2005). By this definition, humans are also keystone predators, with the added responsibility of human consciousness (Hebblewhite et al., 2005). Thus, this chapter repositions humans as fellow predators, who simultaneously possess the capacity to act as stewardship agents in order to reimagine and examine the conflicts in a more inclusive manner.

The Process

My qualitative methodological approach entailed a branching sequence of interrelated methods including participant observations, semi-structured interviews, archival work and wilderness trekking. I used an inductive approach to my research based upon Strauss and Corbin's (1990) grounded theory framework. This framework allowed themes and ideas to emerge throughout my ethnographic fieldwork (conducted from 2013–2015), that informed my subsequent data collection and analyses (Glaser, 1978; Glaser & Strauss, 1967).

For the purposes of this chapter, I focus on one particular method, wilderness trekking. By wilderness trekking, I am referring to my participation with John Davis in the Yellowstone leg of a tri-country journey (Mexico, United States and Canada) during my summer 2013 fieldwork. Known as TrekWest, this endeavour required me to hike and backpack my way across the landscape in order to physically experience it the way that a nonhuman animal would–through traversing the same migratory routes of large predators. My 200-mile wilderness trek from the bigger "cities" of Jackson Hole, Wyoming to Bozeman, Montana yielded over 1600 images, 100 pages of field notes and 33 videos. These data were used to construct a firsthand account of a predator's journey, and they are revealed in this chapter through vignettes. These vignettes are interwoven with theoretical discussions to situate grizzlies, wolves and cougars as equal predatory stakeholders and to illustrate the roads and private real-estate that have made these nonhuman animals' vital habitat into a checkerboard-mosaic of unsafe spaces.

I integrated the stakes of predators using this experiential method as a phenomenological exercise (see Concept Box 6.1). Specifically, I drew upon the theory and methods from Helmreich and Kirksey's (2010) multispecies ethnography to undertake research that was inclusive of the predators in the GYE, whose lives and deaths are linked to human social worlds (Kirksey, 2014). Helmreich and Kirksey (2010) present multispecies ethnography as a mode of inquiry that brings nonhuman animals from the margins into the foreground through transforming methods of anthropological research that were previously restricted to the realms and concerns of humans. A genre concerned with the social and ecological effects of our entanglements with other living beings, this approach alleges that nonhuman animals are more than simply windows and mirrors of symbolic human concerns (Kohn, 2007; Mullin, 1999). Rather, human nature is an interspecies relationship (Tsing, 2009), meaning that our physical, or material, entanglements with other species shape and are shaped by the give and take between humans, nonhuman animals and the environment.

Concept Box 6.1: Phenomenology

Phenomenology provides a framework that scholars in human-animal studies can use to explore the milieu of humans, places and their inhabitants. This concept was borne out of the search to achieve an interconnected and embodied view of beings and their environments, without losing the importance of place, nor succumbing to Cartesian dichotomies [see Jones' Concept Box (Concept Box 4.1) in Chapter 4 of this book].

Phenomenology, as a philosophical paradigm, was based upon the idea that the human experience (of phenomena) is a means by which we can answer questions about the world of which we are not only a part, but also help create. It is the study of how we experience an embodied world laden with tensions between fixed and relational materiality. Experience encompasses not only sensory perception, but also the things we live through and perform (through imagination, thought, emotion and other actions). Thus, phenomenology allows one to examine structures of an embodied being's experience from a first-person point of view.[1]

German philosopher Edmund Husserl (1970) established the paradigm of phenomenology, but it is what came later that is of particular relevance to the concept of dwelling (see Concept Box 6.2 in this chapter). Martin Heidegger (1971) and Maurice Merleau-Ponty (1945) were both concerned with the priorities of Cartesian rationalism that split the mind from the body, nature from culture and beings from their environments, and they deviated from Husserl in his notion that a detached viewpoint of one's experience is possible. These philosophers grounded themselves in the premise that every person [or being] is, before all else, a being-in-the-word. As existential phenomenologists, they expanded Husserl's formulation to argue that an observer cannot separate him or herself from the world. From a phenomenological perspective:

> The world emerges with its properties alongside the emergence of the perceiver in person, against the background of involved activity. Since the person is a being-in-the-world, the coming-into-being of the person is part and parcel of the process of coming-into-being of the world.
>
> (Ingold 2000, p. 168)

The inherent interconnectedness of this inquiry reconciled the limitations of Husserl's detached viewpoint through articulating an understanding of a "being" in-and-of itself; that is, as a human engaged in the making and re-making of the world. Following Heidegger and Merleau-Ponty, a limitation remained of how to translate this philosophy into a method of research that would provide a more nuanced understanding of beings in the world, or agents-in-the-environment. Let us, therefore, try to systematically understand phenomenology through its iteration in Ingold's dwelling perspective (see Concept Box 6.2 in this chapter).

This method that I deployed was an attempt to advocate for a reformed, responsible anthropomorphism that acknowledged the beastly and embodied presence of the predators I was studying in ways that did not simultaneously stifle them (Johnston, 2008), and it looked to address earlier methodological approaches in human-animal studies that were not critical enough of power geometries. During my participation in the trek, I grounded my observations and daily practices in the theoretical framing of Ingold's (1995, 2000) and, later, Johnston's (2008) phenomenological "dwelling perspective" (see Concept Box 6.2). In advocating that an individual learns through engagement, such as trekking, my subsequent analyses reveal a dwelt approach that involved time-deepened and personal encounters rooted in an awareness and appreciation of the inalienable differences across species. By taking this awareness as the point of departure, I argue that only then could a researcher accomplish a relational understanding of nonhuman animals (Johnston, 2008).

Concept Box 6.2: The Dwelling Perspective

Only if we are capable of dwelling, only then can we build (Heidegger 1971, p. 160).

In order to develop a dwelling perspective, Ingold drew upon Heidegger's statement above as a foundation to elaborate more adequate ways of exploring and understanding the relations between humans and the environment, and how those relations differ from those of other embodied beings like nonhuman animals. His dwelling perspective gained its roots from two important influences. The first was Jacob von Uexküll's (1957) concept of the *Umwelt*, who stated that "the world [is] constituted within the specific life activity of the animal" (Ingold, 1995; Knudsen, 1998). He also drew upon ecological psychologist J.J. Gibson's (1979) theory of *affordances*. Affordances, as defined by Gibson (1979), are the properties and possibilities of an object in relation to the observer. For instance, an object such as a pear possesses multiple affordances to a person: it is edible; it can be thrown; it can be used to barter. The potential of an object depends upon the possibilities identified by the actor observing it.

Ingold drew upon these nuanced theories to construct his dwelling perspective and collapse the boundaries between psychology and anthropology. Specifically, he integrated previously separated explorations of cognition and context. He argued that a being's process of cognition, including its acts of thinking, perceiving and learning, must be studied within the ecological context of which that being is a part. Beings create their worlds through the interplay of their entanglements with their surroundings, or their activity of "being" in relation to "context". This activity is what Ingold referred to as *dwelling*. Beings do not just impose meaning into the world because the very world that they are a part of is not a disembodied space (Ingold, 2000; 1995). The contexts within which we dwell are, as Merleau-Ponty stated, a "homeland" to our thoughts.

A dwelt perspective allows scholars to take the philosophical paradigm of phenomenology a step further by providing a methodological underpinning for ethnographic research. Encounters with the world are how we know and construct the very places we seek to describe and understand; that is, we dwell through our lived, embodied *experiences*. When applied methodologically, a dwelt approach involves detailed examination of these experiences and is concerned with an individual's personal perception or account of an object or event. Ethnography, though a broad arena with a multiplicity of practitioners and methods, is largely defined by its effort to examine through immersion, or experience. An active participant records notes, participates in the world that they are studying and uses their own experience to better understand the research topic at hand. Thus, a dwelt phenomenological framework lends itself to a methodological application through ethnography, and opens up new avenues for including the nonhuman animal.

Nonhuman animals' experiences are often defined by the fact that they are not humans. That being said, nonhuman animals are more than just nonhuman beings, they are bodies with matter, that matter, in their own right. An account of human–nonhuman animal relations is incomplete if it focuses only on the human composition of the landscape. Thus, a dwelt phenomenology allows for researchers to consider the interplay of diverse human and nonhuman animal agents in their mutual relations with one another and the environment in which they are participating. When a dwelt phenomenology is employed, a nonhuman animal's experience of the landscape is no longer treated as disembodied relation with the world, just as nonhuman animals are no longer reduced to marginalised representations.

Nature is not "an external, pre-given reality", as Descartes would have one believe. It is always a localised and relational series of phenomena embedded in a particular space and time that is continuously being remade. Humans and nonhuman animals become bounded together in a process of doing. In developing the dwelling perspective, Ingold (2000) provided an integrated concept that opened up new ways to incorporate the complexity of our own relations to the world, as well as the relations of the inhabitants with whom we share the places we dwell.

I integrated Ingold's (2000) and Johnston's (2008) dwelt approach to develop and deploy my phenomenological method of wilderness trekking through predators' habitats in the GYE to critically consider grizzly bears', wolves' and cougars' experiences of the landscape. Exploring human–nonhuman animal relations through a responsible anthropomorphism was in direct contrast to the positivist and objectivist paradigms of most scientific research. Forging accounts of nonhuman animals required me to be attentive to their "beastly" ways through my daily relationships with them, physically and symbolically, which I witnessed firsthand as part of TrekWest. Specifically, my journey revealed the long and varied environmental, political and social history of the flora, fauna and humans that call it home. My wilderness trek and the data it produced helped explain how and why the GYE's predator conflicts unfold, and the theoretical framings that I discuss in the following sections enabled me to contemplate avenues for how we can begin to overcome these conflicts.

When I began my study, I observed that, despite thorough inquiry on the human-carnivore conflicts of the GYE, few scholars had yet to undertake research that de-centred humans and placed them as fellow predators (see Plumwood, 1996 for a notable exception). This gap pointed to the need for innovative research that broke down species boundaries to treat *all* animals as stakeholders. Additionally, few efforts had examined the predator conflicts in a way that integrated theory and methods from multiple disciplines. Fewer still had undertaken this work from a perspective that positioned humans on the same plane as other animals and concurrently as stewardship agents to challenge prevailing science and policy models. My study began to fill this gap through a situated, experiential approach to produce a nuanced understanding of wildlife conflicts, and how they should be studied and addressed.

This chapter examines key findings from my research through attention to the overlapping frameworks of feminist science studies and critical animal geography. It is intended to complement human-animal studies through an interdisciplinary approach that critiques humans' tendency to treat nature and nonhuman animals as external and separate from the influences of human society. Feminist science studies and animal geography are related areas of inquiry that informed the conceptual background of the project through their sophisticated conceptualisations of human–nonhuman animal relations across space, place and time (Urbanik, 2012). Their critiques structured my effort to challenge our fixed ideas about "nature" or "animal" and go beyond traditional models of science and policy (Castree & Braun, 1998, 2001; Haraway, 2008, 2013). Starting from a line of inquiry regarding the complex nature of human and nonhuman animal relations, these interrelated disciplines have helped to bring nonhuman animals out of the proverbial "black box" of nature and into sharper theoretical focus (Urbanik, 2012). The following sections are intended to clarify and elaborate the main concepts of these disciplines to frame my discussions of the predator entanglements of the GYE, such as my first experience traveling into bear country in Grand Teton National Park.

Land Lines

On the day we set off into Grand Teton's backcountry, there had been four recent grizzly bear attacks in the area we intended to hike and camp. I had just bought my first canister of bear spray, and I was still unsure of what I would do if I ever ran into a bear. While I waited on our permits, I began to question myself: "I'm never supposed to run when I see a grizzly. Would I remember not to run? Or, would my fight-or-flight mechanism kick in and cause me to accidentally bolt, triggering the bear's predatory chase instincts and putting my companions in danger?" I sat visualising myself calmly whipping out my bear spray and popping off the cap to deploy it at the right time, when the bear is about 40 feet or so away. I did this over and over, until I convinced myself I had enough imaginary practice and was ready. It had been a few years since I lived in a field station or remote area, and my city safety routines were of little use in the backcountry. Plus, a few weeks prior while safely in my car, I had witnessed the speed at which a grizzly bear can run. I decided to practise a little more outside, popping the cap off and waving the spray canister from the ground up in a zig-zag fashion. After another 15 minutes of practice, a ranger called me over to let me know that he had our permits, and I tucked my bear spray away while we sat in the Flagg Ranch Visitor Center and watched a video about bear safety. Hearing phrases like, "a fed bear is a dead bear", we were instructed on the proper food storage, waste disposal and safety procedures of the Park. While watching, I was reminded of the bear attacks of the previous week where visitors had left out food at their campsite for a sow and her cubs to chance upon. Bear-resistant canister, check. Bear spray, check. Imaginary bear spray deployment practice routine, check. Was I really ready?

Nervousness aside, I thought a lot about that phrase, "a fed bear is a dead bear", as we set off on our hike that day. That phrase is a precautionary heuristic taught to every visitor entering Grand Teton and Yellowstone

National Parks to underscore the importance of safety, for humans and bears. When a bear encounters human refuse, it has found an easy reward. It will then become increasingly aggressive in its efforts to continue to obtain that reward, making the solutions relocation or, too often than not, extermination. Any visitor to the two Parks will attest to the unfathomable beauty of these places. It is the wildlife and the natural features that make the GYE such a marvel (U.S. National Park Service, 2015). However, humans continue to make careless mistakes, for example going to dangerous measures for a photo opportunity. These errors often come at a fatal cost to the lives of nonhuman animals such as predators, which I reflected upon as we worked our way along the String Lake Trail to Leigh Lake (Figure 6.3).

I never needed to use my bear spray, or even pull it out of my holster on that particular trip. Although I avoided any unwanted backcountry encounters, I realised how difficult it is for a nonhuman animal to protect its own safety in those same woods. My early observations followed this logic: for a

Figure 6.3 Photo by Hannah Jaicks-Ollenburger. Part of the Cathedral Group range, as seen from Leigh Lake at sunset.

grizzly bear, cougar or a wolf the journey is unforgiving. A nonhuman animal is not apt to wander through the GYE without accident or risk. Known for its richness and abundance of wildlife diversity afforded by the different habitats of the region, ranging from high alpine forests to dense sagebrush, meadows and geothermal areas (McIntyre & Ellis, 2011; Meagher & Houston, 1998), the GYE also encompasses a staggering complexity of human developments, institutions and practices. Along a predator's journey, it will encounter roads, private real-estate, recreational areas and hunting grounds that are delineated by markers unfamiliar to a nonhuman animal that has never before seen a human.

These dangers are constructed by the overwhelming mix of private entities as well as federal, state and local agencies that dictate what happens to the land and resources. Although Yellowstone and Grand Teton National Parks sit at the heart of this region, providing relief from the threat of hunting or accidentally wandering into the Parks, the invisible lines delineating the safety of the Parks are not inherently known to most humans, let alone a grizzly, cougar or wolf. These contemporary obstacles require attention to the reciprocal relations between humans, wildlife and their local ecosystem (Cronon, 1993, 1996a, 1996b; Lasswell, 1979).

Nature, Animals and the Culture of Domination

To understand these obstacles, I turned to the discipline of feminist science studies. Within this field, scholars have worked to dismantle many of the assumptions implicit in conventional science (Haraway, 1988; Merchant, 1980). This field has distinguished itself through critiques of science's tendency towards "Othering" certain voices and bodies due to the distorted and dichotomising demarcation of a master subject from that of a subordinated object (Barad, 2003; Butler, 1989; Deckha, 2012; Hawkins, 1998; Plumwood, 1993, 2004). Feminist science scholars challenged the Cartesian tendency to hyper-separate the world into dualisms, or dichotomies, like subject/object or nature/culture to reveal the effects of power on the lives of "Others", such as racial, gender and ethnic minorities (Haraway, 1984; Harding, 1998; Keller, 1998). This research has also explored the relationship between power and knowledge, and how the former is used to control and define the latter through scientific practices and principles. For instance, Foucault (1966, 1980) illustrated how information that authorities claim as "knowledge" is a means of social control. I applied Foucault's framework to my research study in order to argue that the prevailing norms in scientific and political processes privilege perspectives that marginalise the needs of the GYE's nonhuman animal "Others".

Feminist science scholars have also challenged many of the essentialist ideas about nature and nonhuman animals (Anderson, 1991; Cronon, 1996a, 1996b; Demeritt, 2004). Alexander Wilson (1991), for example, analysed parks, wildlife conservation projects and media themes to illustrate that "nature" or what is "natural" is not a pre-existent and external entity. Wilson explored the social, political and economic factors that interact and infiltrate our individual histories to shape what we come to identify as "nature". To extend Wilson's argument, I assert that nonhuman animals, like the larger construct of nature, are not external entities that exist separately from human lives.

Additional theorists interested in how nature is produced and constructed have elaborated this critique of a singular and external nature (Katz, 1995, 1998; Merchant, 1980, 2006), which has helped to strengthen the idea that our notion of nature, and nonhuman animals as I argue, form part of historical-ly-contextual processes. These theorists argued against the idea of nature as a stand-alone entity unaffected by, or apart from, human relations. Rather, the material and conceptual separation of human society from nature is what rein-forces and enables capitalist modes of thought about the need to dominate nature (Katz, 1998). This dualism, in which nature and society are categori-cally distinguished from one another, still prevails in thought and practice today. By obscuring the factors underlying our constructs of nature, we authorise new ways to destructively capitalise on the environment (Merchant, 2006; Plumwood, 2004). This dangerously misguided dichotomy of nature and culture has been further critiqued by fellow feminist theorists seeking to repu-diate essentialisms of the often-ignored nonhuman animal "Other" (Anderson, 1995, 1998, 2014; Birke, 2002; Haraway, 1984, 2013; Hawkins, 1998).

I extend these scholars' anti-essentialist critiques of the dualism between humans and nonhuman predators in the GYE to argue that perceived sepa-ration between human and nonhuman animals is a significant reason why current management models are unsuccessful in reducing conflicts across species and preventing further loss of natural resources. Particularly relevant to my study was Hawkins' (1998) expansion of Plumwood's (1993) efforts to restructure hierarchical classification schemes of nonhuman animals by reframing human–nonhuman animal differences as a matter of degree rather than kind. Hawkins employed Charles Darwin's (1859) *Theory of Common Descent* to place humans on an animal continuum and dismantle the anthro-pocentric view of humans as the centre of the universe. Directing attention to predators, according to Hawkins's framework, shifted my focus to a rela-tional schema rather than a categorical or hierarchical one. Hawkins (1998) and Birke (2002) have fostered a biologically-informed view that appreciates both the independent integrity of nonhumans and their continuities with humans. This receptivity has powerful implications for the crisis discipline of conservation because it requires scientists to give up their homogenising and objectifying practices of speciation. It also puts objectivists in the uncomfort-able position of acknowledging the varied and unstable "nature" of nonhu-man animals. This critical construct grounded my effort to expand research methods that do not marginalise the needs of nonhuman animals, without compromising the capacity to conduct an empirical study.

These feminists who sought to dismantle dualistic modes of thought have created a powerful argument for attending to nonhuman animals through a more relational approach. I did not, however, fully comprehend the importance of a relational approach to my research, until I was confronted with the most jarring experience of my journey: a visit to Old Faithful. The juxtaposition between Yellowstone's wilderness that is often considered the "last best place" for wildlife and its adjacency to one of the most popular tourist destinations in the world brought into focus how alarmingly easy it is to push the safety and needs of nonhuman animals to the margins.

Thinking Like a Mountain: Our Conservation Legacy

I was not really sure what I was expecting when I visited Old Faithful for the first time. I had been in the woods for a few days, so my thoughts were mostly about the wild places and wild creatures that had been surrounding me, not on the culture shock I was about to receive. While John and Ed traversed the Winegar Hole Wilderness, an integral part of the Caribou-Targhee National Forest, I was called upon to transfer Ed's equipment and car to the Old Faithful area of Yellowstone. As I drove up, I started to get excited and welcomed a break from my lazy backcountry diet of trail-mix and jerky. Plus, I naively thought, this is my first chance to spend time in one of the most iconic areas of the Park. When I reached the Old Faithful Historic District, I was surprised to find that nearly every space in the megamall-sized parking lot was taken. After seeing no more than a dozen humans in the past two weeks, it was an unexpected surprise to encounter hundreds all at once. If I was feeling overwhelmed, what did that mean for my nonhuman animal counterparts?

While John and Ed got held up with a park ranger to go over their back-country permits, I decided to see for myself if the world-famous geyser is truly as spectacular as it is predictable. Old Faithful, though not the largest geyser in the Park, erupts the most frequently of Yellowstone's big geysers, with an average interval of 90 minutes between eruptions. I looked at the clock on the wall and saw that I only had 15 minutes until the estimated eruption at 1:59 pm, so I went in search of a good vantage point. There were a couple hundred visitors lined up along the designated walkways and, realising how I must have appeared after a week with no shower, I found a semi-isolated place and waited. There were teasers where the geyser would expel a small mist, causing everyone to frantically pull out their cameras, only to be disappointed by the false alarm. At 2:05 pm, Old Faithful started spewing hot water for a solid four minutes. Due to the heavy wind, the 100-foot stream was pushed sideways, giving the appearance of a horizontal waterfall and creating an incredible rainbow in the mist.

After a few minutes, most people seemed conflicted about whether or not to continue filming. Would their clip garner sufficient envy on social media? Instagram and Snapchat were in their infancy at the time, but Facebook and Twitter were already paving the way for people's compulsion to snap, share and seek the "likes" of those not there to witness Old Faithful firsthand. By the end of the eruption, phones were returned to pockets and a steady stream of visitors exited the area to walk back to the parking lot. If ever there was a situation where I witnessed the effects of humans' material and conceptual separation of themselves from nature and wildlife as Cronon (1993), Flores (2001) and Wilson (1991) have described, my afternoon at Old Faithful was it. In that moment, Old Faithful seemed more like a box to check off one's bucket list, rather than a feature on the landscape with intrinsic ecological and sociocultural significance (Katz & Kirby, 1991).

Slightly bemused, I reunited with John and Ed by the car. After repacking our hiking gear with new supplies, we decided to search for milkshakes and shade to escape the heavy sun and smoky skies, a luxury of that a wolf, grizzly or cougar would certainly never get to have.

Making our way into the Old Faithful Inn, I had a chance to see the remarkable architecture of this historic building. We found ourselves inside the more than 110-year-old hotel staring up at its balconies, log-pole framework and crystal-clear view of Old Faithful. After the discovery of a broken milkshake machine in the restaurant, we settled for root-beer floats and returned to take a seat in some rocking chairs scattered around the fireplace in the lounge. Before setting off again on our trek, we discussed the famous hotel and how we felt a bit like cheaters for stealing away indoors for a day with our array of sugary snacks. As anyone who backpacks can attest, there is no greater sensation than your first meal back in "civilization". Similar to my earlier feelings about people's quest to achieve a "natural experience" by witnessing two minutes of Old Faithful's eruption, I was again struck by our collective tendency to treat nature as a separate entity that we must travel to and experience in order to connect with it. I felt entirely hypocritical as I finished off my second root beer float, because I too felt the relief of "separating" myself from nature simply by taking a break from it indoors.

In addition to my own contradictory behaviour, I could not help but observe those surrounding me glued to their phones or computer screens. At that moment, it was hard not to think that our desire to connect with nature only goes as far as we are willing to step outside the confines of our human comforts, myself included. As the largest log hotel in the world, Old Faithful Inn is a place that affords people the luxury and comforts of modern society (e.g., air-conditioning, internet, food), and it juxtaposes this luxury with one of the most phenomenal occurrences of the natural world. We as humans are able to have the intended natural experience and then return to our respective comfort zones. A wolf, grizzly or mountain lion does not ever get to return to its comfort zone because that no longer exists for them. Their worlds are inextricably linked to ours, and their ability to navigate the landscape is often at odds with the wants and needs of humans seeking to live, work and play on that same terrain.

Given the "culture of nature" surrounding places like Old Faithful, it is important make visible how the institutions governing our wildlife and natural resources have the power to shape our views about "nature" and structure our experiences with what is "natural". These decisions contribute to humans' ideas about nature and wildlife as constructs that are separate from society (Flores, 2001; Wilson, 1991), and they exacerbate notions of a "purified" nature uncontaminated by humans to experience in a paradoxically sterilised fashion. Case in point: one can visit the world's most infamous geyser, and then go back inside or to cars parked a mere hundred yards away. This power of institutions to shape dichotomised human-nature relations has persisted throughout America's legacy with conservation over the past 150 years and was readily visible that day by the geyser and inside Old Faithful Inn.

More-Than-Human Animal Geography

To think more about the nature–culture dichotomy in the context of conservation, I engaged with "new animal geography" (Philo & Wilbert, 2000), which I briefly outline here as a means to think with nonhuman animals and critique traditional ways of conceptualising the nature–culture divide. Animal geography is predicated on the assumptions that boundaries between human and nonhuman animals are not fixed and that nonhuman animals are more than just peripheral entities. This sub-field of human geography has distinguished itself by employing feminist science theories to explore the notion that the "who", "when" and "where" a person is in the world shapes that individual's relations with nonhuman animals, which, in turn are heterogeneous and diverse (Emel & Urbanik, 2002). To explain this idea of heterogeneity, researchers position human–nonhuman animal relations as "simultaneously biological, cultural, economic, ethical, geographical and political" (Urbanik, 2012). Animal geography has sought to inform the seemingly mired fields of wildlife conservation and management by integrating awareness of this heterogeneity into policy and practice for constructive outcomes that benefit human *and* nonhuman animals.

Building upon these foundational tenets, researchers like Emel, Wilbert and Wolch (2002) asserted that animal geography has become particularly relevant in light of science's growing recognition of humans' roles in environmental problems. Furthermore, scholars direct attention to the complexity of developing laws that govern human–nonhuman animal relations. As seen in public forums, how we determine and develop practices with respect to nonhuman animals is highly contested across many different stakeholders (Heberlein, 2012). Complicating these contestations has been the increasing acceptance in research that humans have emotional connections, both positive and negative, with nonhuman animals, making it seemingly impossible to deny the interconnectedness of all beings. While earlier scholarship demonstrated the powerful roles that nonhuman animals can play in humans' lives (Herzog & Burghardt, 1988; Rowan, 1988; Shepard, 1978; Shepard & Sanders, 1985), through "new animal geography", which emerged in the mid-1990s, research has exposed a more symmetrical analysis of how processes of power are implicated in human–nonhuman animal relations (Philo & Wilbert, 2000; for a review of how the field of animal geography is expanding, see Buller, 2014, 2015, 2016; Hovorka, 2017, 2018, 2019; Gibbs, 2020a, 2020b).

Whatmore's (2002) concept of hybridity has been useful for re-thinking the reconnections between nature and society. According to Whatmore (2002,

2006), human identities are not created in isolation, but rather they are developed and constantly recreated in relation to animate and inanimate entities. Whatmore, drawing from Latour's (1993, 1999) Actor Network Theory, argued that the actors constituting the world include nonhumans with their own subjective agencies, and that all individuals are entangled in mutually constitutive relations. Finally, a diversity of hybrid relations form and constantly re-form when different human-human, human–nonhuman and non-human–nonhuman animal configurations exist in a given time and place (Urbanik, 2012). The final point made by Whatmore and reiterated by other scholars in the field (2013; Emel, 1995; Hinchliffe, Whatmore, Degan & Kearns, 2005) reminds us that animal geography is also a political project. Researchers in this field have worked hard to ensure that the interests of animals are not unwittingly ignored. Thus, it is a value-laden field that requires researchers to cast judgments about the needs of all actors and how those needs should be met. By documenting and describing my experiences with TrekWest, I was able to explore the entanglements of human and nonhuman predators, which provided me with a chance to take nonhuman animals seriously as active partners.

Lynn's (1998) conceptualisation of "geoethics" has been particularly important as it offered a more nuanced, yet situated, starting point for encountering nonhuman animals. Building on this research, Jones (2000) and Johnston (2008) developed ethical paradigms encouraging scientists and policymakers to see individual nonhuman animals, species and the broader context when conducting research or developing wildlife management protocols. Hence, they instilled an imperative that researchers of human–nonhuman animal relations must incorporate the particular spaces, times and places within which these relations occur.

At the start of my journey, I had not understood any of these ideas well, but I had been rereading Aldo Leopold's *A Sand County Almanac* and working through some of my trekking observations as we reached the end of our trip. One passage in particular stuck with me:

> All ethics so far evolved rest upon a single premise: that the individual is a member of a community of interdependent parts. His [*sic.*] instincts prompt him to compete for his place in that community, but his ethics prompt him to also co-operate (perhaps in order that there may be a place to compete for). The land ethic simply enlarges the boundaries of the community to include soils, waters, plants and animals, or collectively: the land.
>
> (Leopold 1949, p. 239)

Leopold understood the necessity of enlarging the boundaries around which we develop our ethics, particularly as they relate to decisions about the environment and its inhabitants. However, on my trek, I arrived at the question of *how* we achieve this inclusivity. As my journey came to a close, this lingering question took me back to my graduate school library and in search of paradigms to ground truth these observations and structure them in a way that would enable others to undertake similar work. Drawing and building upon feminist science studies and animal geography, as I have done here, allowed me to achieve this goal, and the methodological approach I describe in this chapter helped put theory into practice. What I present here is a platform for the reader to consider ways in which they can do the same.

Parting Ways

On the last part of my trek, we hiked into the Centennial Mountains of Montana and Idaho along the Odell Creek Trail and the Continental Divide trail. As we entered the backcountry, we followed a small mountain stream under a thick forest canopy that eventually led us into a bowl-like valley with the rugged Centennials towering over us on both sides. We had spent the previous night swapping stories at Lillian Lake, and I learned more background about the heavily contested U.S. Sheep Experiment Station that we were on our way to scope out. Our target was to reach the grounds of this research station that spans nearly 28,000 acres. The station is in remote and critical habitat for imperilled species like grizzly bears, wolves, bighorn sheep, lynx and wolverines. It also puts domestic sheep and immigrant workers directly in harm's way of migrating carnivores, which has led to the death of multiple grizzlies in the past decade (Gilman, 2012). Due to the ecological threats posed by its location and the ostensible financial insolvency of the station, environmental organisations have been lobbying for its closure. Its operation, as with other matters related to predator management, continues to be an ongoing subject of litigation.

Over the course of our hike, we found numerous piles of fresh bear scat that underscored the ever-present danger of running into a grizzly. Despite having spent a lot more time in grizzly bear habitat by that point, I still felt the need to practise my bear spray preparedness. If and when I ran into a bear, I wanted to have my muscle memory and instincts as finely tuned as possible. Eight miles in and no bear sightings besides scat and scratched trees, I was surprised when we reached the station. The only demarcation of this station was a wooden sign laden with carvings (see Figure 6.4). It was the sole indicator that we had left the trail and entered the grounds of a major agricultural research base. This invisible line reiterated my earlier observation that the boundaries we use to denote land ownership and land use have no intrinsic or symbolic bearing to a grizzly, wolf or cougar. However, the material consequences for nonhuman animals that cross onto the station lands can be, and often are, fatal. After the miles I had hiked, my last stop at the station crucially underscored the many obstacles nonhuman animals face as we ask them to adhere to boundaries and rules delineated by humans, many of which we have trouble deciphering as well.

Predators and other nonhuman animals, as my trek revealed, are repeatedly tasked with seemingly insurmountable challenges of successfully crossing highways, avoiding catastrophe by not eating livestock or attacking a

Figure 6.4 Photo by Hannah Jaicks-Ollenburger. The U.S. Sheep Experiment Station
          lands overlap with critical grizzly bear ranges in the Greater Yellowstone
          Ecosystem, and the station is demarcated by this wooden sign.

person and navigating an unforgiving climate. Undertaking the trek also clarified the complex issues that humans must confront to live and survive in the region. Humans, like nonhuman predators, face material and social obstacles that directly affect their lives and livelihoods, and there is a sharp feeling of futility that I witnessed amongst residents with regards to the decision-making processes of the GYE. The socio-environmental history of the GYE has produced a political system that constrains human and nonhuman predators' ability to survive and coexist on the landscape. Consequently, the conflicts over the management of the GYE's large carnivores remain an ongoing problem that will only continue to get worse as we face a changing global climate and an economic system that devalues and pushes the humans and wildlife of the GYE further into the margins.

When we reached my final destination where I would wish John and Ed a safe journey for the remaining month of their trek, we spent that last morning riding bikes along the dirt road leading out of the Centennials. As we said our goodbyes, I asked John what he thought about possible solutions to these many conservation challenges of protecting wildlife, natural resources and our own human interests. He stood for a second, and then responded:

I think that we should always tell the ecological truth. We should always let the scientists tell us clearly what would be best for wild nature and wildlife. Start with that information…but you [also] have to start talking. You must listen to the concerns and the needs of the various interest groups…of the landowners, the government officials, the conservationists and the animals themselves. And too often it's going to be a power struggle. But you have to do it.

(personal communication, August 2013)

Like Leopold and many others before him, John reaffirmed my own belief that science and policy are powerful tools for managing pressing conservation issues, but without a sense of inclusivity and awareness of power dynamics, these tools are insufficient. Trekking, coupled with a critical theoretical framework, allowed me to overcome some of these limitations.

Conclusions

Discussion of my trek as a predator's journey through the GYE re-evokes the primary question that continues drives my research in present day: What is one to make of the predator conflicts that manifest themselves so visibly on the landscape? The GYE is far more than an outlined area on a map. It is a *place*, one filled with humans, wildlife and ecological features of economic, political, sociocultural and biological significance (Flores, 2001). A powerful symbol as the heart of the American West, this terrain is where conflicts over land use, conservation, federal power and private ownership unfold. Its multi-faceted social and ecological features permit many ways of knowing and valuing the landscape (Tuan, 1974).

These different ways of experiencing, knowing, and appreciating the GYE are all equally valid. Problematically, different ways of use and valuation play out in public arenas as deeply divisive conflicts over preservation versus conservation, non-use versus wise-use, federal governance versus state-hood ideology, public and private land ownership and many others. I have repeatedly witnessed these conflicts imbue the debates over predator management in this area with symbolic meanings far beyond the physical presence of a wolf, grizzly or cougar.

Thus, the methodological discussion in this chapter acknowledges the multiplicity of predatory stakeholders in the GYE to highlight the complex relationships between humans and their surroundings. Further, it identifies the ways that nonhuman animals like predators are often relegated to being vehicles for humans' political or social agendas and advances a phenomenological framing for how to overcome this marginalisation. Through interweaving theory and an experiential depiction of predators' paths through the GYE, I draw attention to the subtle ways that humans and wildlife are dependent upon and embedded in their environment. Further, by examining how nature, humans and nonhuman animals are all agents actively participating in the GYE, I destabilise humans as the sole agents in control of history.

Destabilising these processes, in turn, captures the give and take, the action and reaction, of human and nonhuman animal relationships (and conflicts) situated within the context of the Greater Yellowstone Ecosystem. It is through phenomenological exploration of these interactions within and across species, couched in a critical theoretical framework that allows for greater understanding of (and ability to address) the predator management conflicts that have emerged and become cemented over time.

My journey through the region was only a brief glimpse into what the GYE's large carnivores encounter on a daily basis, yet it served to underscore the serious challenges that nonhuman predators as well as humans must confront in the region. My wilderness trek illustrated firsthand that addressing these conflicts requires constant reform, localised approaches and continued evaluation of our conservation efforts if we hope to preserve the diversity of wildlife, resources and places that make all of our lives richer and more worthwhile. My discussion is therefore intended to inform science and policy on methods we can take to more effectively and proactively mediate conflicts and manage humans and nonhuman animals' needs as well as a theoretical rationale for why this approach is essential. As fellow predators, we depend upon the diversity of wildlife and natural resources in our places of dwelling for survival; as humans, we have the capacity and responsibility to take more inclusive, process-oriented and integrative approaches to protecting this diversity – for ourselves and our fellow carnivores – to ensure our coexistence long into the future.

Note

1 Early philosophers did not describe experience in terms of a "being" as I do, but rather they used the term "person". To extend this program of study into the realm of animal geography, I substituted the term "being", as it more readily encompasses nonhuman animals as actors with their own subjective experiences in and of the world.

References

Anderson, J. E. (1991) 'A conceptual framework for evaluating and quantifying naturalness', *Conservation Biology*, 5(3): 347–352.
Barad, K. (2003) 'Posthumanist performativity: toward an understanding of how matter comes to matter', *Journal of Women in Culture and Society*, 28(3): 801–831.
Berger, J. (1991) 'Greater Yellowstone's native ungulates: myths and realities', *Conservation Biology*, 5(3): 353–363.
Berger, J. (2009) *The better to eat you with: fear in the animal world*, University of Chicago Press, Chicago, IL.
Birke, L. (2002) 'Intimate familiarities? Feminism and human-animal studies', *Society and Animals*, 10(4): 429–436.
Blanchard, B. M. & Knight, R. R. (1991) 'Movements of Yellowstone grizzly bears', *Biological Conservation*, 58(1): 41–67.
Buller, H. J. (2014) 'Animal Geographies I', *Progress in Human Geography*, 38(2): 308–318.

Buller, H. J. (2015) 'Animal geographies II: methods', *Progress in Human Geography*, 39(3): 374–384.

Buller, H. J. (2016) 'Animal geographies III: ethics', *Progress in Human Geography*, 40(3): 422–430.

Butler, J. (1989) 'Foucault and the paradox of bodily inscriptions', *Journal of Philosophy*, 86(11): 601–607.

Castree, N. & Braun, B. (1998) 'The construction of nature and the nature of construction', in B Braun & N Castree (eds.), *Remaking reality: Nature at the end of the millennium*, Routledge, London, pp. 3–42.

Castree, N & Braun, B. (2001) *Social nature: theory, practice, and politics*, Blackwell Publishing, Oxford.

Cronon, W. (1993) 'The uses of environmental history', *Environmental History Review*, 17(3): 1–22.

Cronon, W. (1996a) 'The trouble with wilderness: or, getting back to the wrong nature', *Environmental History*, 1(3):7–28.

Cronon, W. (1996b) *Uncommon ground: rethinking the human place in nature*, WW Norton and Company, New York, NY.

Darwin, C. (1859) *On the origin of species by means of natural selection, or the preservation of favoured races in the struggle for life*, John Murray, London.

Deckha, M. (2012) 'Toward a postcolonial, posthumanist feminist theory: centralizing race and culture in feminist work on nonhuman animals', *Hypatia*, 27(3): 527–545.

Demeritt, D. (2004) 'What is the 'social construction of nature'? A typology and sympathetic critique', *Progress in Human Geography*, 26(6): 767–790.

Emel, J. (1995) 'Are you man enough, big and bad enough? Ecofeminism and wolf eradication in the USA', *Environment and Planning D*, 13: 707–734.

Emel, J. & Urbanik, J. (2002) 'Animal geographies: exploring the spaces and places of human–animal encounters', *Society and Animals*, 10(4): 407–412.

Emel, J., Wilbert, C. & Wolch, J. (2002) 'Animal geographies', *Society and Animals*, 10(4): 407–412.

Flores, D. (2001) *The natural West: Environmental history in the Great Plains and Rocky Mountains*, University of Oklahoma Press, Norman, OK.

Foucault, M. (1966) *Les mots et les choses*, Éditions Gallimard, Paris.

Foucault, M. (1980) *Power/knowledge: selected interviews and other writings, 1972–1977*, Pantheon, New York, NY.

Gibbs, L. M. (2020a) 'Animal geographies I: hearing the cry and extending beyond', *Progress in Human Geography*, 44(4): 769–777.

Gibbs, L. M. (2020b) 'Animal geographies II: killing and caring (in times of crisis)', *Progress in Human Geography*, 45(2): 371–381.

Gibson, J. J. (1979) *The ecological approach to visual perception*, Houghton Mifflin, Boston, MA.

Gilman, S. (2012) 'Growing grizzly population conflicts with USDA Sheep Station', *High Country News*, 44(3). https://www.hcn.org/issues/44.3/growing-grizzly-population-sparks-conflict-with-a-usda-sheep-research-station

Glaser, B. G. (1978) *Theoretical sensitivity*, Sociological Press, Mill Valley, CA.

Glaser, B. G. & Strauss, A. L. (1967) *Discovery of grounded theory: strategies for qualitative research*, Aldine, Chicago, IL.

Hannibal, M. H. (2012) *Spine of the continent: the most ambitious wildlife conservation project ever undertaken*, Lyons Press, Guilford, CT.

Haraway, D. J. (1988) 'Situated knowledges: the science question in feminism and the privilege of partial perspective', *Feminist Studies*, 14(3): 575–599.

Haraway, D. J. (2008) *When species meet*, University of Minnesota Press, Minneapolis, MN.

Haraway, D. J. (2013) *Simians, cyborgs, and women: the reinvention of nature*, Routledge, New York, NY.

Harding, S. (1998) 'Rethinking standpoint epistemology: what is 'strong objectivity'?', in EF Keller & HE Longino (eds.), *Feminism and science*, Oxford University Press, Oxford, pp. 264–279.

Hawkins, R. Z. (1998) 'Ecofeminism and nonhumans: continuity, difference, dualism, and domination', *Hypatia*, 13(1): 158–197.

Hebblewhite, M., White, C. A., Nietvelt, C. G., McKenzie, J. A., Hurd, T. E., Fryxell, J. M. & Paquet, P. C. (2005) 'Human activity mediates a trophic cascade caused by wolves', *Ecology*, 86(8): 2135–2144.

Heberlein, T. A. (2012) *Navigating environmental attitudes*, Oxford University Press, Oxford.

Heidegger, H. (1971) *Poetry, language, thought*, trans. A Hofstadter, Harper and Row, New York, NY.

Heinen, J. T. (2007) 'Coexisting with large carnivores: lessons from Greater Yellowstone', *Practice*, 9(2): 140–141.

Helmreich, S. & Kirksey, S. E. (2010) 'The emergence of multispecies ethnography', *Cultural Anthropology*, 25: 545–576.

Herzog, H. A. & Burghardt, G. M. (1988) 'Attitudes toward animals: origins and diversity', in AN Rowan (ed.), *Animals and humans sharing the world*, University Press of New England, Hanover, NH, pp. 75–94.

Hinchliffe, S, Whatmore, S, Degan, M & Kearns, M. (2005) 'Urban wild things: a cosmopolitical experiment', *Journal of Environment and Planning D: Society and Space*, 23: 643–658.

Hovorka, A. J. (2017): Animal geographies I: globalizing and decolonizing. *Progress in Human Geography* 41(3): 382–394.

Hovorka, A. J. (2018): Animal geographies II: Hybridizing. *Progress in Human Geography*, 42(3): 453–462.

Hovorka, A. J. (2019) Animal geographies III: species relations of power. *Progress in Human Geography* 43(4): 749–757.

Husserl, E (1970) *The crisis of European sciences and transcendental phenomenology: an introduction to phenomenological philosophy*, Northwestern University Press, Chicago, IL.

Ingold, T (1995) 'Building, dwelling, living: how animals and humans make themselves at home in the world', in M Strathern (ed.), *Shifting contexts: transformations in anthropological knowledge*, Routledge, London, pp. 57–80.

Ingold, T. (2000) *The perception of the environment: essays on livelihood, dwelling and skill*, Routledge, New York, NY.

Jobes, P. C. (1991) 'The Greater Yellowstone social system', *Conservation Biology*, 5(3): 387–394.

Jobes, P. C. (1993) 'Population and social characteristics in the Greater Yellowstone Ecosystem', *Society & Natural Resources*, 6(2): 149–163.

Johnston, C. (2008) 'Beyond the clearing: towards a dwelt animal geography', *Progress in Human Geography*, 32(5): 633–649.

Jones, O. (2000) '(Un)ethical geographies of human—non-human relations: encounters, collectives and spaces', in C Philo & C Wilbert (eds.), *Animal spaces, beastly places: new geographies of human-animal relations*, Routledge, New York, NY, pp. 268–291.

Katz, C. (1995) 'Under the falling sky: apocalyptic environmentalism and the production of nature', in A Callari, S Cullenberg & C Biewener (eds.), *Marxism in the postmodern age: confronting the New World Order*, Guildford, New York, NY, pp. 276–282.

Katz, C. (1998) 'Whose nature, whose culture? Private productions of space and the preservation of nature', In B Braun & N Castree (eds.), *Remaking reality: nature at the end of the millennium*, Routledge, London, pp. 46–63.

Katz, C. & Kirby, A. (1991) 'In the nature of things: the environment and everyday life', *Transactions of the Institute of British Geographers*, 16(3): 259–271.

Keller, E. F. (ed.) (1998) *Feminism and science*, Oxford University Press, Oxford.

Kirksey, E. (ed.) (2014) *The multispecies salon*, Duke University Press, Durham, NC.

Kohn, E. (2007) 'Animal masters and the ecological embedding of history among the Ávila Runa of Ecuador', *Anthropological Perspectives*, 106–129.

Lasswell, H. D. (1979) *The signature of power: buildings, communication, and policy*, Transaction Books, New Brunswick, NJ.

Latour, B. (1993) *The pasteurization of France*, Harvard University Press, Cambridge, MA.

Latour, B. (1999) *Pandora's hope: essays on the reality of science studies*, Harvard University Press, Cambridge, MA.

Leopold, A. (1949) *A sand county almanac*, Oxford University Press, New York, NY.

Lynn, W. (1998) 'Animals, ethics, and geography', in J Wolch & J Emel (eds.), *Animal geographies: place, politics, and identity in the nature-culture borderlands*, Verso, London, pp. 280–297.

McIntyre, C. L. & Ellis, C. (2011) *Landscape dynamics in the Greater Yellowstone Area*, National Resource Technical Report no. 506, National Park Service, Fort Collins, CO.

Meagher, M. & Houston, D. B. (1998) *Yellowstone and the biology of time: photographs across a century*, University of Oklahoma Press, Norman, OK.

Merchant, C. (1980) *The death of nature: women, ecology, and the Scientific Revolution*, Harper Collins, New York, NY.

Merchant, C. (2006) 'The scientific revolution and the death of nature', *Isis*, 97: 513–533.

Merleau-Ponty, M. (1945) *Phenomenology of perception*, Éditions Gallimard, Paris.

Mullin, M. (1999) Mirrors and windows: sociocultural studies of human-animal relationships. *Annual Review of Anthropology*, 28: 01–224.

Philo, C. & Wilbert, C. (eds.) (2000) *Animal spaces, beastly places: new geographies of human-animal relations*, Routledge, London.

Plumwood, V. (1993) *Feminism and the mastery of nature*, Routledge, London.

Plumwood, V. (1996) 'Being prey', *Terra Nova*, 3: 32–44. Print.

Plumwood, V. (2004) 'Gender, eco-feminism, and the environment', in R White (ed.), *Controversies in environmental sociology*, Cambridge University Press, Cambridge, MA, pp. 43–60.

Rasker, R. & Hansen, A. (2000) 'Natural amenities and population growth in the Greater Yellowstone region', *Human Ecology Review*, 7(2): 30–40.

Rowan, A. N. (1988) 'Introduction: the power of the animal symbol and its implications', in AN Rowan (ed.), *Animals and humans sharing the world*, University Press of New England, Hanover, NH, pp. 1–12.

Shepard, P. (1978) *Thinking animals: animals and the development of human intelligence*, Viking Press, New York, NY.

Shepard, P. & Sanders, B. (1985) *The sacred paw: the bear in nature, myth, and literature*, Viking Penguin Incorporated, New York, NY.

132 Hannah Jaicks-Ollenburger

Smith, D. & Ferguson, G. (2005) *Decade of the wolf: returning the wild to Yellowstone*, Rowman & Littlefield, Lanham, MD.

Strauss, A. L. & Corbin, J. (1990) *Basics of qualitative research: grounded theory procedures and techniques*, Sage, Newbury Park, CA.

Tsing, A. (2009) 'Supply chains and the human condition', *Rethinking Marxism*, 21(2): 148–176.

Tuan, Y. F. (1974) *Topophilia: a study of environmental perception, attitudes, and values*, Prentice Hall, Inc., New York, NY.

von Uexküll, J. (1957) 'A stroll through the worlds of animals and men: a picture book of invisible worlds', in CJ Schiller (ed.), *Instinctive behavior: the development of a modern concept*, International Universities Press, Madison, WI, pp. 5–80.

Urbanik, J. (2012) *Placing animals: an introduction to the geography of human-animal relations*, Rowman & Littlefield, Lanham, MD.

U.S. National Park Service (2015) *National Park Service Overview*, viewed 25 February 2015, http://www.nps.gov/news/upload/NPS-Overview-2015-update-2-25.pdf

US National Park Service (2019) *Greater Yellowstone Ecosystem*, map, 1: 3,168,000, US National Park Service, viewed 29 August 2019, https://www.nps.gov/yell/learn/nature/greater-yellowstone-ecosystem.htm

Whatmore, S. (2002) *Hybrid geographies: natures, cultures, spaces*, Sage, London.

Whatmore, S. (2006) 'Materialist returns: practising cultural geography in and for a more-than-human world', *Cultural Geographies*, 13(4): 600–609.

White, P. J., Garrott, R. A. & Plumb, G. E. (eds.) (2013) *Yellowstone's wildlife in transition*, Harvard University Press, Cambridge, MA.

Wilson, A. (1991) *The culture of nature: North American landscape from Disney to the Exxon Valdez*, Between the Lines, Toronto.

7 How to do multispecies-ethnographies when exploring human-(wild) animal interactions

Affect, multisensory communication and materiality

Susan Boonman-Berson and Séverine van Bommel

Introduction

"The members of the N!ore community wanted elephants to be included as stakeholders in the co-management of the area", says social scientist Neil Powell as he shares[1] his experience researching co-management in Kaokoland, Namibia. He explains:

> According to the N!ore community members, elephants participate in the planning and harvesting of Marula trees (Schlerocarya caffra) in the area. The fruits of the Marula trees in the area are eaten both by the elephants and by the N!ore people. Therefore, the community members have decided together with the elephants which trees are 'people trees' to be used by people and which trees are 'elephant trees' to be used by elephants. According to the N!ore, if 'elephant trees' are depleted by the people, the elephants become angry and they will destroy all the 'people trees' in the area in revenge. Therefore, the N!ore felt that the elephants were stakeholders too in the management of the area and they asked for the elephants to be included in the multi-stakeholder negotiation process.

The question this human-elephant example brings forward is "how do we deal with these non-human participants in multispecies research"? This question is particularly relevant in many human-wildlife conflict cases worldwide.

The human-elephant example also clearly shows that the boundaries drawn between humans and wild animals are not as universal as we tend to think in western societies. It raises the issue that interpretive, qualitative researchers can be at a loss when confronted with a fieldwork context that challenges the taken-for-granted anthropocentrism embedded in our research methods and methodology (Kohn, 2007). Conventionally, we – the authors of this chapter – have been trained to investigate what animals mean to humans. However, due to the current worldwide increase in human-wild animal conflicts (Knight, 2000; WWF, 2006) and the current progress in demonstrating animal intelligence (Bradshaw, 2017; De Waal, 2016), many suggest that the focus of research should shift to try to understand how humans and (wild) animals

DOI: 10.4324/9781351018623-9

co-create meaning in interaction. For instance, a number of fieldwork experiences such as the one described above have led us to question the human exceptionalism in most research and research methodology (Haraway, 2008a, 2008b). Our own experiences have led us to increasingly understand human-wildlife conflicts as an evolving social relationship between two different entities, of which only one happens to be human. Based on these experiences, we have increasingly realised that animals have equal agency and intersubjectivity to human actors. Therefore, we argue that by ignoring considerations about animal agency in qualitative, interpretive research, we will continue to place animals at the margins of research practices, thereby inevitably becoming complicit in their silencing and making them invisible in research accounts. In this chapter we propose the opposite.

Over the last few years, we have reflected on the question, "How do we include animals in our research more symmetrically"? Specifically, how should we *do* this methodologically? Based on ethnographic research we know that empathy is important in understanding and bridging the lifeworld of "the Other", who can be very different from us (Rogers, 1961). Additionally, research shows us that animals are intelligent, sentient beings and have emotions and feelings sometimes very similar to human animals (Bradshaw, 2017; De Waal, 2009; Hurn, 2012). This suggests that empathy can also be extended to animals. Thinking through the idea of mutuality in empathy allows us, humans, to achieve a certain understanding of animals' lived experiences and thus, theoretically, animals can be included as research participants in fieldwork. This is what sparks our interest in multispecies ethnographies, a branch of ethnographic work which has specifically taken up the challenge to decentre humans in research methodology to create a more egalitarian discursive space in and out of the field (Buller, 2015; Hodgetts & Lorimer, 2015; Kirksey & Helmreich, 2010).

However, in practice, such an approach also presents a number of methodological concerns. Here, we discuss three concerns: The first is that animals, apparently, cannot figure as participants in our research projects and methods in the same way that people can because animals do not speak or write the way we do. This means that they cannot be interviewed or join focus group discussions. It also means that we cannot check our interpretations with our animal research participants in the same way that we would in ethnographic fieldwork (Hamilton & Taylor, 2017). This raises a second concern about ethical implications, as we cannot ask animals for informed consent. Therefore, it is difficult to know whether animals are willing or not to be part of our research. If animals cannot speak for themselves then it is either the human participants in our research who speak for the animals or it is "us", the researchers, who speak for "them". The former puts us squarely back into the realm of humanism because then we are merely investigating the meaning(s) that animals have for humans, instead of taking animals seriously as research participants in their own right, both as individuals and collectives. To include animals as participants in social science research raises the question how this would make "us" as social scientists any different from ethologists interpreting animal behaviour. This brings us to our third, and last, concern, namely that speaking for animals includes the risk of anthropomorphising animals or attributing "uniquely human features to animals" (Daston & Mitman, 2005, p. 2). In animal ethology, ascribing humanity to animals is a taboo and an absolute "no go". This mode of thought seems to have crossed over into other fields of research.[2] Researchers with this stance hold on to the underlying assumption that humans and animals are so different that people cannot, and accordingly should not, assume to understand the lived experiences of animals and, thus, humans cannot speak for animals (Hamilton & Taylor, 2017). Consequently, the researchers who hold these views tend to approach animals as static and generically defined by species. These three methodological concerns point to some of the challenges of interspecies communication, which leaves us with the puzzle of how to "do" a multispecies ethnography in practice.

In this chapter we tackle this question by first discussing the affective, multisensory, material and relational aspects of doing multispecies ethnographies (see the Concept Box in this chapter). We then draw on three small case studies from our own empirical work on management of wild boars, black bears and elephants to identify three interconnected methodological strategies that researchers can use in practice to overcome the challenges of interspecies communication. We call these strategies "engaging with affect", "multisensory reading and writing", and "bringing back materiality".

The animal turn in interpretive research

In most social science literature, scholars emphasise the ways in which humans experience the world rather than accounting for and including the experience of nonhumans (Urbanik, 2012). This approach is increasingly challenged by posthumanist scholars who claim that humans are no longer the only subjects that matter in research. By accepting the relational status of multispecies encounters, a "less fixedly human and more dynamic approach to boundaries" might be revealed (Buller, 2014, p. 7). This perspective acknowledges that ideas about what it means to be "human" emerge in interaction with nonhumans. This relational approach, including the intention to approach humans and (wild) animals more symmetrically, initiated a call for an "animal turn" in social sciences (cf. Buller, 2014).

With the animal turn, a variety of concepts have invited us to re-think animals and humans as active beings that co-constitute and co-create the world in interaction (cf. Lulka, 2009). These notions include that of "multispecies ethnography"[3] (Buller, 2015; Hamilton & Placas, 2011; Kirksey & Helmreich, 2010; Lorimer & Srinivasan, 2013), which we use in this chapter to problematise the predominance of solely human intentionality and agency and re-think human-animal relations in terms of mutuality.

Multispecies ethnography draws on a variety of academic fields such as more-than-human geography, philosophy and science and technology studies (Ogden et al., 2013). Much of the scholarly work that includes multispecies ethnographies consists of studies about interactions between humans and domestic animals (see Power, 2008 on human-dog interactions, and Maurstad et al., 2013 on humans and horses). Neo and Ngiam (2014), for example, discuss that dolphin trainers are able to recognise and isolate specific, spontaneous, "invented" new behaviours by dolphins in captivity. The trainers then reinforce dolphins' behaviour by teaching them to perform in the presence of human trainers. Trainers recognising new behaviours by the dolphins, and dolphins learning to understand what the trainer intends to teach them, is only possible through developing an affective relationship between trainer and dolphin. Essentially, trainer and dolphin attune to each other. Additionally, Neo and Ngiam argue that the dolphins show their "dolphin-ness" as part of their relationship with humans: the new behaviour of dolphins is not entirely shaped by humans (it is invented by the dolphins), nor it is entirely produced by only the dolphins (the behaviour is most likely only performed in captivity and subsequently reinforced by human trainers) (2014).

When it comes to human-wildlife interactions, multispecies ethnographies face some specific methodological challenges compared to those that focus on people and domestic animals. Rarely have wild animals been treated as ethnographic subjects by researchers "hanging out" with them for extended periods of time. Only few exceptions come to mind, such as the work of Jane Goodall with chimpanzees in Gombe (2009), the work of Cynthia Moss with wild elephants in Amboselli National Park (1982) and the work of Shaun Ellis with wolves in the Nez Percé Indian lands in Idaho (2009).[4] Mostly, the complex nature of human-wild animal interactions does not lend itself easily to every day and sustained relationships (as in domestic and livelihood contexts); rather these interactions are dynamic, intermittent and fleeting (Bear & Eden, 2008, 2011) which makes it even harder to include animals as ethnographic subjects in research.

However, although methods vary, scholarship on human-domestic animal and human-wild animal interactions has drawn attention to *affect, multisensory communication*[5] and *materiality* (see Concept Box 7.1) as conceptual tools to study interactions between humans and animals (cf. Boonman-Berson et al., 2016, 2019; Candea, 2013; Haraway, 2008a). For instance, several scholars have found the notion of *affect* – in terms of intersubjectivity (Archambault, 2016; Latimer & Miele, 2013) – helpful to capture the embodied nature of multispecies ethnographies (see the chapter by Jones in this book). Blanco et al. (2015), for example, have found the notion of affect helpful to capture the active exchange of affects between salmon and people on Chilean fish farms. Lorimer (2008) has explored the way in which counting corncrakes involves various kinds of embodied skills and emotions. Other scholars have used the notion of *multisensory communication*, drawing on insights from Deleuze and Guattari (1988), to describe the signs (visual, olfactory, auditory, tactile) materialised in words, signals or things, that are communicated between humans and wild animals through the writing and reading of these signs in the landscape. For example, Hinchliffe et al. (2005) refer to "the footprints and other marks that water voles leave or make as they go about their business" (p. 647) to describe the signs left by water voles which are subsequently "read" by humans. Herrero (2002) recounts the way a radio-collared, male black bear tried to throw off the researcher following him by "reading" the traces the researcher left in the landscape. This example, in contrast to the water voles' study, draws attention to the mutuality involved in the back-and-forth communication between humans and wild animals (Boonman-Berson et al., 2016).[6] And, finally, some scholars, whose works are inspired by Donna Haraway, have drawn attention to the role that *materiality* plays in mediating human-wild animal relationships. As an example of the creation of a "material reality" between humans and wild animals, Barua (2014) elaborates how human-wildlife conflicts in India were mediated by alcohol. The alcohol consumed by villagers while guarding their fields was the exact substance that attracted the elephants to their villages. Although the concepts of affect, multisensory communication and materiality are used by researchers studying human-wildlife interactions, it remains unclear how they went about this methodologically (see Figure 7.1). Drawing on our own experience, we will discuss how we approached this in our own work and reflect on its methodological implications. We do this by way of three different cases. The first describes the affective learning processes of wildlife managers, wild boars and the researcher at the Veluwe, the Netherlands. The second case zooms in on the role of multisensory communication in black bear management in Colorado, USA. The third case unpacks the negotiations between wild African elephants and people in a human-wildlife conflict in Laikipia, Kenya. We conclude this chapter with discussing the three methodological strategies we have employed and then return to the concerns of doing multispecies ethnography raised in the introduction.

Engaging with affect in wild boar management practices

"You need to think along with the boars", a wildlife manager explains to me[7] while we follow a wild boar trail. We traced the trail by observing wild boar rooting alongside it, and evidence of some boar hair at a tree adjacent to the trail. This wild boar manager said that he knew where "his" wild boars roamed. To know and manage these wild boars, affective relationships (see concept box in this chapter) are crucial. In fact, without an affective aptitude it is nearly impossible to count and hunt these boars – which are the two key elements of wild boar management at the Veluwe (Koninkrijksrelaties, 2012, 2017). More specifically, it becomes possible to collect the required data, or hunt wild boars, by relying on such affective relations and informal ways of knowing.

Additionally, the wild boars learn to tune in and learn from their environment, including humans. As a result, to be able to count and hunt the wild boars, wildlife managers need to learn to be affected by wild boars. For instance, at the Veluwe, in order to count and hunt wild boars, the managers often make use of fixed wild observation posts. This means that it is particularly important to attract the boars to these areas. This, in turn, requires an intuitive way of knowing and particular skills to communicate with the boars. That became clear to me, for instance, at the end of an entire evening of counting wild boars. At the time, around ten o'clock at night, we wanted to leave our observation post. However, at that particular moment, a wild boar family was roaming around the area and feeding on the corn that the wildlife manager had strewn earlier that evening. Suddenly, the wild boar manager started making a short farting sound, effectively telling the wild boars to leave the spot. He produced this sound several times before one of the female boars responded by uttering a short grunt. She and the other boar family members then walked a few meters to the edge of the feeding area, away from us. They continued feeding at the edge; however, the female boar that had responded with the grunt was still alert, with her head and ears lifted. The wildlife manager repeated the farting sound one more time and, subsequently, this female turned around and, quietly, walked away into the forest. The other boar family members then followed suit, leaving us in the distance. On our way back out of the forest, the wildlife manager explained that he some-times uses a roaring sound instead. He decides and experiments on the spot what sound to use to communicate with the wild boars. To become affected by the wild boars and their wider ecology, he had spent extended periods of time observing and experimenting with multisensory forms of communica-tion. In the case of wild boars, these forms are principally non-visual. Similarly, the boars at the Veluwe learn to become affected by their wider environment, including the wildlife managers making particular sounds. On the occasion described, it appeared that the female boar had learned to inter-pret the sound as unfamiliar, but not threatening. As a result, the boars did not run away. This, in turn, prevented them from creating a negative associa-tion with this area, which is important since the same spots are used for hunt-ing purposes. Indeed, ecological research indicates that wild boars have site-specific awareness (Morelle et al., 2014).

However, to understand and grasp these affective learning processes between managers and wild boars required me, as a researcher, to move between the formal scientific approaches of investigating humans and ani-mals and adopt an open approach to grasp their "affective experiences" (Barua & Sinha, 2017). Consequently, I had to learn to be affected by the wild boars and their wider environment. To illustrate this affective learning process, I offer an anecdote that contributed to this learning process. It relates to one of the most important things in counting wild boars: to become unnoticed by them. While sight is the predominant sense for humans, wild boars are, in addition to sight, much more dependent on their senses of hearing and of taste. Therefore, not being seen, heard and smelled by wild boars is important; otherwise, they would never come to the spot where they can be counted. One way to be "unseen" is to dress with dark colours, use no artificial odours such as perfume, avoid pungent sandwich ingredients, refrain from squeaky sand-wich bags and avoid rustling clothes. And most important, be silent. As a nov-ice doing this fieldwork, I was familiar with these rules, although on my first field-day counting wild boars (about five o'clock in the evening), I brought

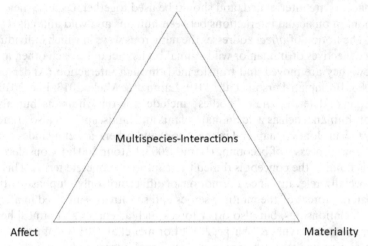

Figure 7.1 The three key-concepts and their relations that are necessary to understand the variety of multispecies interactions in human-animal research. All three concepts zoom in on actual interactions between humans and wild animals.

Source: Produced by Susan Boonman-Berson and Séverine van Bommel

sandwiches in a bag. Overwhelmed with the silence at the spot, I had to decide to either not eat the sandwiches and ignore the sound of my stomach or take my sandwiches out of the bag and possibly cause some noise. With a lot of patience, I took the sandwiches out as silently as possible. However, my fellow counter and I did not know if this act would influence the – possible, not yet visible – animals around us. Reflecting on such field experiences has helped me become better attuned to and engaged with both the wild boars and the human participants involved. Specifically, it taught me to respect and adapt to the code of conduct in the practice of wild boar management. By being immersed in the field, "getting to know" the wild boar participants and the area they live in, as well as the human-participants, has contributed to the generation of detailed affective knowledge about the human-wild boar interactions at the Veluwe. Doing so can uncover details taken for granted by the experienced wildlife researcher and/or manager, contribute to creating a mutual affective understanding between humans and wild boars and, possibly, reveal other non-lethal ways of managing the boars.

Concept Box 7.1: Affect, multisensory communication and materiality

In human-wildlife interactions, the interrelated concepts of *affect*, *materiality* and *multisensory communication* specifically draw attention to exchanges and mutuality involved in these interactions. These three concepts are interrelated, and should be used together, because they all zoom in on actual interactions between humans and wild animals.

The notion of *affect* addresses the numerous ways in which individual or collectives of human or wild animal bodies act upon each other, and how they are moved and transformed through interaction (Anderson, 2006; Boonman-Berson et al., 2019; Latimer & Miele, 2013; Pile, 2010). In general terms, these "bodies" include not only humans but also non-humans such as wild animals, plants and landscapes. As also argued by Owain Jones (Chapter 4 in this book) and others, affect includes continuous processes of becoming (Barad, 2003; Latour, 2004a). Considering this fluidity, the concept is difficult to capture in concrete terms. This is especially relevant since the notion of affect not only emphasises the material forces of the multiple senses and sensitivities involved in affective relationships, but also other forces such as changes in atmosphere for example (Hynes & Sharpe, 2015; Lorimer et al., 2017). Affect is thus co-constituted through human-(wild) animal interactions.

While affect is co-created through human-(wild)animal interactions, the notion of *multisensory communication* helps us understand the processes taking place in these interactions. Specifically, multisensory communication focuses on the various senses, besides the visual, used in these processes. When dealing with human-(wild) animal communication through signs (visual, olfactory, auditory, tactile), we can only access and take into account the (limited) material aspects of these communication processes. However, we can still grasp something of the lifeworld of wild animals by taking material aspects into account, since these can reveal "hidden messages" (Gooch, 2008; Marvin, 2005). The notion of multisensory communication can be further specified by the notion of multi-sensory writing and reading (Boonman-Berson et al., 2016, p. 72), which "recognizes that both humans and animals leave traces as well as trace, interpret and respond to these traces". For instance, humans write with words, but also "write" by means of, for instance, placing fences, planting particular crops and using specific scent or auditory signals. Animals, in turn, interpret and enact these human "writings". This becomes observable by changes in their behaviour and movement. Animals "write" by means of signals or things, such as footprints, feces, left fur, particular sounds and scents. As such, the notion of multisensory communication highlights the continuous process of ongoing interactions and interpretations enacted by humans and (wild) animals.

The notion of *materiality* is interrelated with the idea of multisensory communication. This concept particularly addresses the creation of a "material reality" mediating the communication process between humans and (wild) animals. Material artefacts can contribute to this communication process, such as technical/non-living materials (e.g.

camera traps, audio equipment, the use of drones, rocks), but also "living-materials" (e.g. bees in communicating space between humans and elephants, foxhounds as interpreters between humans and foxes). All these materialities contribute to understanding the ongoing and mutual processes between humans and (wild) animals.

Multisensory writing and reading in managing human-black bear interactions

"And I got a call from her [a local resident] one morning and she said, 'I made a terrible mistake'". A local "Bear Aware" volunteer[8] recounts a situation that he regularly encounters in the area where he lives–the Colorado Front Range of the Rocky Mountains, USA. Here, both the total number of human inhabitants and black bear inhabitants have been increasing since the 1990s (Baruch-Mordo, 2007; Center of the American West, 2001; Colorado State Demography Office, 2014). Simultaneously, recorded conflicts between humans and black bears have been increasing (Vieira, 2011). Black bear management in the Front Range is confronted with these conflicts and has chosen a variety of strategies to co-exist with the bears. Part of this management involves the deployment of Bear Aware volunteers who, together with wildlife managers, deal with local human-bear conflicts. On this occasion, the caller explained that she had made a quick trip to the supermarket but had left the garage door open. The woman ran a catering business from her home and therefore she had a large freezer full of wedding cakes, shrimp and other "goldmine food" for bears. When she got back home, she discovered a mother black bear and cub in her freezer. The lady said to the volunteer that she will never leave her garage door open again. However, the volunteer replied that "it might be too late". Indeed, early next morning the woman called back. But this time the bears had torn the siding off her garage to get back in. The volunteer explained to me: "When they [bears] learn something like that you know, if they get rewarded like that with that kind of food they're gonna be back". Subsequently, he helped her by moving the chest freezer inside her house and spraying the garage with ammonia to remove any odours attractive to the black bears. The scent of ammonia signals the bears to "stay away". Additionally, they left the siding torn off and the garage door open. Thus, the black bears could smell that something had changed, interpret that message, and could see that the chest freezer was gone. This time the woman took no further risk, and she was really careful about keeping the entry way to her home clean of any food smells, wiped everything down with ammonia, and kept the freezer in the living room of her house for the rest of the year. The effect of this multisensory communication strategy implemented by the woman was that the bears passed her home one more time, but never entered the garage again. This example also illustrates that communicating with wild animals, such as bears, requires thinking through a variety of senses: "just" using a (visual) look is not enough, as it is the smell that is highly alluring to the bears.

A wildlife manager explained to me that *"each bear has its own strategy"*. He said that we – humans – need to understand a particular bear's strategy in order to communicate with that specific bear. For instance, not all bears tear off wall sidings. He recalled that a bear popped out window screens, went in and headed straight to the refrigerator, ate all the contents, and went right back out the same way (s)he came in. That happened every day, usually between 5:30 and 7:30 in the morning, when people were still asleep or in their bedrooms. The bear entered about 27 different houses, every time using the same strategy. This bear had learned to read the typical way humans arrange their homes–that each home has a refrigerator with delicious food for bears– and (s)he had learned to pop out window screens and enter the houses. Wildlife managers in the Front Range intend to go along with this typical individual bear-behaviour. They learn to read the writings of these black bears and translate their interpretations into management strategies. The implemented management strategy results in (multisensory) writing directed towards a particular bear and, as a result, the strategy used is generally aimed at the continuation of humans and bears sharing the same area without conflicts. Examples of the employed strategies involve the use of rubber bullets, bean bags, noise, the use of particular odours, etc. These strategies are implemented until the bear goes away. The strategy chosen depends on the situation and on the specific bear. In cases where human safety is of concern and a bear has become "corrupt", such as the one breaking into homes day after day, the 2-strike management directive is implemented (although the results of this policy are highly questioned among wildlife managers), and multisensory communication is used. First, the bear is attracted into a bear-trap, for instance, by using the same kind of fruit that the bear had been eating. Once the bear has been trapped, (s)he will be ear tagged and relocated to prime bear habitat. The ear tag is used to determine if a particular bear causes problems a second time. In that case, the policy requires killing the bear. In practice, the majority of wildlife managers intend to avoid implementation of this 2-strike directive by gathering detailed information about the "problematic" interactions and simultaneously seeking the best possible solution for both humans and black bears. This means that they experiment with a variety of strategies that include multisensory forms of communication, both directed at the involved humans and individual black bears. Examples include teams of "Bear-Aware volunteers" educating humans on how to live with bears, or wildlife managers enforcing bear-proof garbage and storage rules. Wildlife managers also educate bears by removing or locking any attractant, or by using odours, sounds and material artefacts such as rubber bullets. In essence, these strategies include different forms of multisensory writing and reading between humans and black bears. This results in humans and black bears co-shaping the landscape in which both dwell.

Elephants and people negotiating space in Laikipia, Kenya

"You have to negotiate with elephants to keep them inside the fenced areas". That is what the Kenyan Wildlife Service officer told me[i] in the Laikipia area

of Kenya. We were looking at an electric fence. The wires were broken, and the fence was destroyed. The fence had been constructed to reduce human-wildlife conflicts, especially crop-raiding by elephants which is a significant problem in areas where people and elephants share the same space. In Laikipia, electric fences have been built in an attempt to find a technical solution to this problem. Fences are material artefacts and a spatial means of controlling the interactions between people and elephants by dividing the landscape into areas for elephants, and areas for cultivation. The officer explained that as soon as the fences had been constructed, the elephants quickly learned to break them. The elephants discovered that their tusks do not conduct electricity. Fence-breaking elephants – especially the habitual raiders – learned to pull and push the wires of an electric fence with the tops of their tusks until the wires snapped. This presented the Kenya Wildlife Service officers with a problem. To prevent the elephants from breaking the wires the wildlife managers removed 2/3 of the tusks of the fence-breaking elephants. They left 1/3 of the tusk to make sure that the central nerve was still protected.

However, this did not stop the elephants from breaking the wires: "They now take a young male and they throw him into the fence. He starts thrashing around, breaking the fence. And that is how they get out", said the officer. Having lived and worked with the elephants for a long time, the Kenyan Wildlife Service officers had learned to identify different herds of elephants and different individuals. They also had learned to trace the migration of the elephants and how the elephants negotiated the landscape full of human activity. Affect between people and elephants is an integral part of the management of the landscape in Laikipia. People are affected by the elephants when the elephants break the fences and go crop raiding on neighbouring farms. The elephants are also affected by the fences that people construct to keep them inside a certain area and out of other areas, thereby restricting their movement and access to water and food. The fences can be seen as a form of communication with the elephants. The elephants read the landscape and respond by either respecting the fences or by breaking them. People and elephants in the Laikipia area thus negotiate with each other to share space and resources.

After a stop to investigate the broken electrical wires, we all got back into the car and continued our journey. The wildlife manager proceeded to tell his story: "We are now also experimenting with beehive fences". I was puzzled and I asked him what beehive fences were. He explained that beehive fences are made of hives that are suspended every ten meters and linked together with an interconnecting non-electric wire. These fences are constructed around villages and agricultural fields to keep the elephants out. "So instead of fencing the elephants in, we are now fencing people and crops in", he explained. The beehive fences are constructed in such a way so that should an elephant touch one of the hives or the interconnecting wire, then all the beehives along the fence begin to swing and release the bees. Elephants are afraid of bees, which can sting them inside their trunks, and thus elephants flee when they hear them buzzing (King et al., 2007). The idea of beehive fences was developed by Lucy King, an ethologist and researcher, who discovered that elephants have a natural fear of bees. She started working in Laikipia in 2008 by experimenting with beehive fences and monitoring the responses of the elephants to these fences (King et al., 2009, 2011). In one of her interviews, she reflected on her interactions with the elephants:

> What's amazing is that we can just walk and see what's happening with the elephants as they approach. If we're lucky, we get one of our collared elephants approaching the beehive fences. So high and low technology, it suddenly combines, and we can look at the moving technology of this elephant hitting a very simple cheap beehive fence, and we can learn a lot from these interactions. We've just collared 20 more elephants in our study area hoping to get much more data on the actual movements of collared elephants around our project's site.[9]

Lucy King is thus able to read the collared elephants' responses to the beehive fences through the movement/spatial data translated in numbers and dots on a computer screen. King has developed a finely tuned ability to interpret the elephant behaviour that she reads from the data produced by material artefacts such as collars and remote cameras. The remote cameras allow her to see when the elephants become nervous and back off from the fence. The technology of the beehive fences, collars and remote cameras have an impact on the interactions between elephants and people. The elephants learn to stay out of the agricultural fields, and people allow the elephants to share the area where they both live. During our interview, King explained how some elephants are notorious crop raiders: "They are like naughty children". She went on to describe how these elephants needed to be disciplined to behave like "proper wild elephants", and how she considered it part of her job to do that. According to King, this is important because if the elephants do not show a proper fear or respect for the beehive fences, it would be impossible for humans and elephants to co-exist in the area.

This case illustrates intensive interactions between wildlife managers, local farmers, researchers, elephants, different types of fences, bees and collars. It shows how different actors participate in new relationships, and how elephants and people communicate and negotiate the use of space and resources to co-exist in the Laikipia area in Kenya. This communication and negotiation are mediated by material artefacts such as electric fences, beehives, collars, computer screens and remote video cameras. Both people and elephant play an active part in the conflict and in the negotiation process. Finally, this example points to how people and elephants reach an agreement regarding the use of shared resources such as water, vegetation and land, thus shaping an unfolding material reality of co-existence.

Doing multispecies ethnographies

In this chapter we have showed that *affect, multisensory writing and reading*, and *materiality* are key concepts to methodologically bridge the human-(wild) animal divide. We have demonstrated how such an approach is useful for enabling interactions between humans and wild animals that promote mutual learning. Importantly, as the case studies have demonstrated, these three concepts are interrelated. More specifically, understanding processes of multisensory "writing" and "reading" between humans and (wild) animals is not possible without taking in consideration the affective relationships emerging from their encounters. Additionally, humans and (wild) animals become affected through materiality, which in turn contributes to finding ways for human-animal mutual understanding and decision making. Next, we zoom in on the three interconnected methodological strategies, which we have illustrated by discussing the examples, and we reflect on the three concerns we raised in the introduction of this chapter.

The first methodological strategy we discuss is *"engaging with affect"*. The cases studies have shown various ways in which humans and wild animals affect each other through particular encounters. In the case of the wild boars, humans and wild boars learned to be affected by counting and hunting areas, and by each other. In the case of the black bears, wildlife managers learned to be affected by individual black bears through findings ways to communicate with them, and black bears learned to be affected by learning where (not) to dwell. As for the beehive fences in Laikipia, Kenya, humans, elephants and bees engaged in affective relations of conflict and negotiation, which ultimately involved happy people, angry bees and fearful elephants. The notion of affect allowed us to focus on the subjective agency and affective life worlds of both humans and wild animals. The people in our case studies (including us researchers) talked about encounters with particular individual animals and how they had experienced those encounters as profoundly affective experiences. In a similar way, the behaviour of the wild animals showed how they were also affected by the people with whom they shared their landscape. By describing the affective encounters between humans and wild animals, we were able to make visible the new ways of inter-being and relating that developed through their interactions.[10]

The second methodological strategy we discuss is *"multisensory reading and writing"*. All the brief excerpts of the case studies have demonstrated that humans and wild animals communicate with each other, negotiate and seem to deliberate (cf. Driessen, 2014; Meijer, 2017). In the case of wild boars, managers communicate with the boars to attract them to particular sites using corn and specific sounds intended to deter boars without developing negative associations with that site. In the case of the black bears, both bears and humans negotiated space by interpreting the various signs left by each other and, subsequently, responding to them in their own human/bear way. In the case of the elephants, humans communicated with elephants by leaving various kinds of fences in the landscape as signs not to trespass. The elephants communicated back by respecting the fences or by breaking them. In all cases, the way humans and wild animals communicate is not limited by human language but involves all senses. All participants in the communication process (humans or animals) write and read in their own specific way.

They differ by which principal sense is used to communicate and navigate their surroundings, such as wild boars who seem to navigate their environment using magnetic fields (Morelle et al., 2014). Additionally, participants differ in how they translate the knowledge gained from the interactions. This might be translated for humans into written texts such as specific policy plans, or among elephants who use a special "bee rumble" for warning other elephants, sometimes kilometres away, to stay away from a certain area that contains angry bees (King et al., 2010). Although it is not an easy task to translate this multisensory communication into (human) words, it is not impossible. It requires investigating the ongoing communication between humans and wild animals as expressed by the varied multisensory processes at play.

The third methodological strategy we discuss is to *"bring back materiality"*. The elephant case-study has drawn particular attention to the use of radio-collars and camera traps to further our understanding of the interactions between humans and wild animals, and also to communicate with wild animals. Fences are material artefacts that contribute to our communication of "space" to specific wild animals. However, when material artefacts are introduced as a way of communicating with, and understanding wild animals, this involves political and ethical considerations. For instance, it requires thinking about the meaning that the wild animals in question make of these material artefacts. Are these material artefacts introduced as a way of acting *upon* the wild animals (such as hunting wild boar to prevent damage to human properties) or as a way of acting *with* the wild animals (such as the beehives as an intermediary device to communicate between both parties)? The use of material artefacts in managing human-wildlife interactions raises ethical questions. For instance, what is the impact of radio-collared versus non-radio collared elephants on their well-being and interactions with conspecifics? What about the impact of translocating a black bear to another bear's territory and/or the role of ear tags for bear life? We argue that the inclusion of materiality in thinking about managing human-wildlife interactions requires thinking about new ways of introducing and designing animal-specific material artefacts.

We now return to the three concerns raised in the introduction. The main issues at stake are the subjectivity involved in investigating human-animal interactions, how to deal with that methodologically and, relatedly, the symmetrical account we aim to portray through our methods. First and foremost, we acknowledge it is far from an easy task to translate affective and multisensory communication into (human) words, let alone grasp them by means of scientific research methods. Based on our experiences, we argue that this can be achieved by using a combination of two research strategies: on the one hand, conducting ongoing, long-term participant-observations and, on the other hand, doing joint research projects with people who have found ways to understand the wild animals under investigation. The latter research strategy might involve people, like wildlife managers, who have been in the field for a long time and have detailed knowledge and established long term relationships with the wild animals under investigation. This second strategy might also involve initiating joint research projects with ethologists and (behaviour) ecologists (cf. Mason & Hope, 2014), who often make use of technology to understand a particular wild animal. The two described research strategies involve the use of and experimentation with multisensory forms of communication, for example the case of using materials such as ammonia. Additionally, these strategies requires the researcher him/herself to be reflective of his/her affective relations with both the animals and the humans under investigation. Investigating human-animal interactions, then, becomes an ongoing learning process, driven by being affected by the animals and humans involved. In doing so, such a research strategy is a joint effort between the involved humans and (wild) animals (Barua & Sinha, 2017; Boonman-Berson, 2018; Van Dooren & Rose, 2012).

There are two concerns that remain. First, the ethical concern that animals cannot be consulted in the same way as people to either join or reject participation in a research project. Our results suggest – but still need further investigation – that animals can be "consulted" by establishing and reflecting on the affective relationships between humans (researchers) and wild animals. The second concern that remains is about the problems and risks of objectifying animals while conducting joint research projects with ethologists. Based on our experiences, we propose a third research strategy to engage with affect; namely, through the use of technology. The latter is a challenging strategy because it runs the risk of objectifying the animal. However, as we have argued, material artefacts could be considered as a means of understanding and relating with each other, and be conducive to cohabitation as a co-operative undertaking. Furthermore, as Bradshaw (2017) states, to study trans-species behaviour, which includes human-animal research, disciplines such as ethology and sociology have to merge. This suggests that a new social-ethological vocabulary still needs to be established to investigate human-animal interactions.

Considering the above suggested research strategies in doing multispecies-ethnographies, we need "promiscuous research methods" (Swanson 2017, p. 83) that transcend the boundaries between natural sciences and social sciences. In the case of negotiating with elephants, for example, this included radio collaring elephant by Lucy King to read their responses to the beehive fences. This resonates with Hodgetts and Lorimer (2015) who have suggested that exploring interactions between humans and more-than-human others requires tracking and data collection devices affixed to difficult-to-follow animals. Although it may seem like a radical suggestion to have social scientists working with natural science data collection methods, social scientists have always been learning methods from other disciplines to carry out their research. Most of us may not be formally trained in archival research, but we learn methods from history if we need to (Swanson, 2017). These sorts of collaborations are transcending earlier approaches in studies of science in which the practices of "knowledge making" of natural scientists were being put under the microscope by social scientists. Multispecies ethnography shows that more-than-human others are not simply brought into being through the enactment and performative practices of natural scientists. More-than-human others are active participants in the construction of reality. As multispecies ethnography challenges the nature-culture divide, it is not surprising that it would also challenge the way this divide has been translated into rigid disciplinary boundaries. Natural scientists are increasingly becoming "critical friends" of multispecies ethnographers (Van Dooren et al., 2016) through collaboration in research and writing (Buchanan et al., 2015; Kirksey et al., 2016; Lestel et al., 2006). So, we may instead have to think about doing "multispecies-*ethnologies*" that combine tools from ethnography and ethology to better understand human-(wild) animal interactions. It is necessary to include and connect to ethological accounts of investigating animals in the conventionally known methods of *doing* ethnography (see chapter by Maderson & Adams-Elsner in this volume). More specifically, using the notion of "multispecies-*ethnologies*" involves re-thinking *and* aligning knowledge from the natural sciences and social sciences.

To conclude, affective ways of knowing in human-animal research can be strengthened by reciprocally engaging more with ethological research accounts (with their writings and through joint research projects), as well as with the local practices of wildlife management. How such multispecies-ethnologies are done in practice is an entirely empirical matter, as there can be many strategies according to the context where the research is conducted. We have presented three methodological strategies that contribute to new ways of investigating human-animal interactions, especially in pursuing a more democratic and egalitarian way of including animals into research.

Notes

1 The second author of this Chapter.
2 This is a discussion that keeps coming up when we present our work at various research seminars.
3 Other concepts include zoopolis (Donaldson & Kymlicka, 2011; Wolch, 1998), the dwelling perspective (Ingold, 2000, 2005), hybridity (Latour, 2007; Whatmore, 2002), responsible anthropomorphism (Johnston, 2008), multispecies social practices (see Bear, 2011; Buller, 2015; Eden & Bear, 2011; Law & Mol, 2008), lively biogeographies (Barua, 2014; Lorimer, 2010), and multinaturality (Brettell, 2016; Latour, 2004b; Lorimer, 2012).
4 It needs to be noted that to our knowledge none of these researchers have ever used the term "multispecies ethnography" to characterise their work.
5 Multisensory communication is in this chapter elaborated by the notion of multisensory writing and reading.
6 Specifically, this kind of communication emphasises the intimately entangled ongoing character of these interactions between human and wild animal. Additionally, the spatiality of these interactions shape the landscape in which they are situated (Boonman-Berson, 2018).
7 The first author of this Chapter.
8 Neighbours talking to neighbours about how to live with black bears.
9 Read the full interview at https://www.paulallen.com/innovator-profiles-lucy-king-uses-bees-to-keep-the-peace-between-humans-and-elephants/#2k0MBcaIzg Feild7.99

10 This is in line with findings by Van Dooren and Rose (2012) about the making of storied places by humans, penguins and flying foxes. "Storied places" refers to the way in which some specific animals narrate, in whatever way, their specific places. It highlights the way in which places are understood and embedded in broader histories and systems of meaning, including the stories and experiences generated by the animals involved (Boonman-Berson et al., 2019).

References

Anderson, B. (2006) Becoming and Being Hopeful: Towards a Theory of Affect. *Environment and Planning D: Society and Space* 24, 733–752. https://doi.org/10.1068/d393t

Archambault, J.S. (2016) Taking Love Seriously in Human-Plant Relations in Mozambique: Toward an Anthropology of Affective Encounters. *Cultural Anthropology* 31, 244–271. https://doi.org/10.14506/ca31.2.05

Barad, K. (2003) Posthumanist Performativity: Toward an Understanding of How Matter Comes to Matter. *Signs: Journal of Women in Culture and Society* 28, 801–831. https://doi.org/10.1086/345321

Barua, M. (2014) Bio-Geo-Graphy: Landscape, Dwelling, and the Political Ecology of Human-Elephant Relations. *Environment and Planning D: Society and Space* 32, 915–934. https://doi.org/10.1068/d4213

Barua, M., Sinha, A. (2017) Animating the Urban: An Ethological and Geographical Conversation. *Social & Cultural Geography* 1–21. https://doi.org/10.1080/14649365.2017.1409908

Baruch-Mordo, S. (2007) *Black Bear-Human Conflicts in Colorado: Spatiotemporal Patterns and Predictors (MSc.)*. Colorado State University, Fort Collins, Colorado.

Bear, C. (2011) Being Angelica? Exploring Individual Animal Geographies. *Area* 43, 297–304. https://doi.org/10.1111/j.1475-4762.2011.01019.x

Bear, C., Eden, S. (2008) Making Space for Fish: The Regional, Network and Fluid Spaces of Fisheries Certification. *Social & Cultural Geography* 9, 487–504. https://doi.org/10.1080/14649360802224358

Bear, C., Eden, S. (2011) Thinking like a Fish? Engaging with Nonhuman Difference through Recreational Angling. *Environment and Planning D: Society and Space* 29, 336–352. https://doi.org/10.1068/d1810

Blanco, G., Arce, A., Fisher, E. (2015) Becoming A Region, Becoming Global, Becoming Imperceptible: Territorialising Salmon in Chilean Patagonia. *Journal of Rural Studies* 42, 179–190. https://doi.org/10.1016/j.jrurstud.2015.10.007

Boonman-Berson, S. (2018) *Rethinking Wildlife Management: Living With Wild Animals*. Wageningen University, Wageningen, the Netherlands. https://doi.org/10.18174/455279

Boonman-Berson, S., Driessen, C., Turnhout, E. (2019) Managing Wild Minds: From Control by Numbers to a Multinatural Approach in Wild Boar Management in the Veluwe, the Netherlands. *Transactions of the Institute of British Geographers* 44, 2–15. https://doi.org/10.1111/tran.12269

Boonman-Berson, S.H., Turnhout, E., Carolan, M. (2016) Common Sensing: Human-Black Bear Cohabitation Practices in Colorado. *Geoforum* 74, 192–201.

Bradshaw, G.A. (2017) *Carnivore Minds: Who these Fearsome Animals Really Are*. Yale University Press, New Haven & London.

Brettell, J. (2016) Exploring the Multinatural: Mobilising Affect at the Red Kite Feeding Grounds, Bwlch Nant yr Arian. *Cultural Geographies* 23, 281–300. https://doi.org/10.1177/1474474015575472

Buchanan, B., Chrulew, M., Bussolini, J. (2015) On Asking the Right Questions. *Angelaki* 20, 165–178. https://doi.org/10.1080/0969725X.2015.1039821

Buller, H. (2014) Animal Geographies I. *Progress in Human Geography* 38, 308–318. https://doi.org/10.1177/0309132513479295

Buller, H. (2015) Animal Geographies II: Methods. *Progress in Human Geography* 39, 374–384. https://doi.org/10.1177/0309132514527401

Candea, M. (2013) Habituating Meerkats and Redescribing Animal Behaviour Science. *Theory, Culture & Society* 30, 105–128. https://doi.org/10.1177/0263276413501204

Center of the American West (2001) Growth Data Sheet [WWW Document]. URL http://www.centerwest.org/futures/archive/data_sheet.html

Colorado State Demography Office (2014) Population Totals for Colorado and Sub-state Regions [WWW Document]. URL http://www.colorado.gov

Daston, L., Mitman, G. (Eds.) (2005) *Thinking with Animals: New Perspectives on Anthropomorphism*. Columbia University Press, New York.

De Waal, F. (2016) *Are We Smart Enough to Know How Smart Animals Are?* W. W. Norton & Company, New York.

De Waal, F.B.M. (2009) Darwin's Last Laugh. *Nature* 460, 175. https://doi.org/10.1038/460175a

Deleuze, G., Guattari, F. (1988) *A Thousand Plateaus: Capitalism and schizophrenia*. University of Minnesota Press, Minneapolis; London.

Donaldson, S., Kymlicka, W. (2011) *Zoopolis: A Political Theory of Animal Rights*. Oxford University Press, Oxford; New York.

Driessen, C. (2014) *Animal Deliberation: The Co-evolution of Technology and Ethics on the Farm*. Wageningen University, Wageningen, the Netherlands.

Eden, S., Bear, C. (2011) Reading the River Through 'Watercraft': Environmental Engagement through Knowledge and Practice in Freshwater Angling. *Cultural Geographies* 18, 297–314. https://doi.org/10.1177/1474474010384913

Ellis, S. (2009) *The man who lives with wolves*. Three Rivers Press, New York.

Gooch, P. (2008) Feet following Hooves, in: Ingold, T., Vergunst, J.L. (Eds.), *Anthropological Studies of Creativity and Perception*. Ashgate Publishing Limited, Surrey, pp. 67–80.

Goodall, J. (2009) *In the Shadow of Man*. Mariner Books, Boston.

Hamilton, J.A., Placas, A.J. (2011) Anthropology Becoming …? The 2010 Sociocultural Anthropology Year in Review. *American Anthropologist* 113, 246–261. https://doi.org/10.1111/j.1548-1433.2011.01328.x

Hamilton, L., Taylor, N. (2017) *Ethnography after Humanism: Power, Politics and Method in Multi-Species Research*. Springer, London.

Haraway, D. (2008a) Otherworldly Conversations, Terran Topics, Local Terms, in: Alaimo, S., Hekman, S. (Eds.), *Material Feminisms*. Indiana University Press, Bloomington, pp. 157–187.

Haraway, D. (2008b) *When Species Meet*. University of Minnesota Press, Minneapolis.

Herrero, S. (2002) *Bear Attacks: Their Causes and Avoidance*, Rev. ed Lyons Press, Guilford, Conn.

Hinchliffe, S., Kearnes, M.B., Degen, M., Whatmore, S. (2005) Urban Wild Things: A Cosmopolitical Experiment. *Environment and Planning D: Society and Space* 23, 643–658. https://doi.org/10.1068/d351t

Hodgetts, T., Lorimer, J. (2015) Methodologies for Animals' Geographies: Cultures, Communication and Genomics. *Cultural Geographies* 22, 285–295. https://doi.org/10.1177/1474474014525114

Hurn, S. (2012) *Humans and Other Animals: Cross-Cultural Perspectives on Human-Animal Interactions*. Pluto Press, London.

Hynes, M., Sharpe, S. (2015) Affect. *Angelaki* 20, 115–129. https://doi.org/10.1080/09 69725X.2015.1065129

Ingold, T. (2000) *The Perception of the Environment: Essays on Livelihood, Dwelling and Skill*. Routledge, London.

Ingold, T., (2005) Epilogue: Towards a Politics of Dwelling. *Conservation and Society* 3, 501–508.

Johnston, C. (2008) Beyond the Clearing: Towards a Dwelt Animal Geography. *Progress in Human Geography* 32, 633–649.https://doi.org/10.1177/0309132508089825

King, L.E., Douglas-Hamilton, I., Vollrath, F. (2007) African Elephants Run from the Sound of Disturbed Bees. *Current Biology* 17, R832–R833. https://doi.org/10.1016/j.cub.2007.07.038

King, L.E., Douglas-Hamilton, I., Vollrath, F. (2011) Beehive Fences as Effective Deterrents for Crop-Raiding Elephants: Field Trials in Northern Kenya. *African Journal of Ecology* 49, 431–439. https://doi.org/10.1111/j.1365-2028.2011.01275.x

King, L.E., Lawrence, A., Douglas-Hamilton, I., Vollrath, F. (2009) Beehive Fence Deters Crop-Raiding Elephants. *African Journal of Ecology* 47, 131–137. https://doi.org/10.1111/j.1365-2028.2009.01114.x

King, L.E., Soltis, J., Douglas-Hamilton, I., Savage, A., Vollrath, F. (2010) Bee Threat Elicits Alarm Call in African Elephants. *PLOS ONE* 5, e10346. https://doi.org/10.1371/journal.pone.0010346

Kirksey, E., Hannah, D., Lotterman, C., Moore, L.J. (2016) The Xenopus Pregnancy Test: A Performative Experiment. *Environmental Humanities* 8, 37–56. https://doi.org/10.1215/22011919-3527713

Kirksey, S.E., Helmreich, S. (2010) The Emergence of Multispecies Ethnography. *Cultural Anthropology* 25, 545–576. https://doi.org/10.1111/j.1548-1360.2010.01069.x

Knight, J. (Ed.) (2000) *Natural Enemies: People-Wildlife Conflicts in Anthropological Perspective*. Routledge, London; New York.

Kohn, E. (2007) How Dogs Dream: Amazonian Natures and the Politics of Transspecies Engagement. *American Ethnologist* 34, 3–24. https://doi.org/10.1525/ae.2007.34.1.3

Koninkrijksrelaties, M. Van B.Z. en (2012) Flora- en faunawet.

Koninkrijksrelaties, M. Van B.Z. en (2017) Wet natuurbescherming.

Latimer, J., Miele, M. (2013) Naturecultures? Science, Affect and the Non-human. *Theory, Culture & Society* 30, 5–31. https://doi.org/10.1177/0263276413502088

Latour, B. (2004a) How to Talk About the Body? The Normative Dimension of Science Studies. *Body & Society* 10, 205–229. https://doi.org/10.1177/1357034X04042943

Latour, B. (2004b) *Politics of Nature: How to Bring the Sciences into Democracy*. Harvard University Press, Cambridge, MA.

Latour, B. (2007) *Reassembling the Social: An Introduction to Actor-Network-Theory, Clarendon Lectures in Management Studies*. Oxford University Press, Oxford.

Law, J., Mol, A. (2008) The Actor-Enacted: Cumbrian Sheep in 2001, in: Knappett, C., Malafouris, L. (Eds.), *Material Agency: Towards a Non-Anthropocentric Approach*. Springer US, Boston, MA, pp. 57–77. https://doi.org/10.1007/978-0-387-74711-8_4

Lestel, D., Brunois, F., Gaunet, F. (2006) Etho-Ethnology and Ethno-Ethology. *Social Science Information* 45, 155–177. https://doi.org/10.1177/0539018406063633

Lorimer, J. (2008) Counting Corncrakes: The Affective Science of the UK Corncrake Census. Social Studies of Science 38, 377–405.

Lorimer, J. (2010) Elephants as Companion Species: The Lively Biogeographies of Asian Elephant Conservation in Sri Lanka. *Transactions of the Institute of British Geographers* 35, 491–506. https://doi.org/10.1111/j.1475-5661.2010.00395.x

Lorimer, J. (2012) Multinatural Geographies for the Anthropocene. *Progress in Human Geography* 36, 593–612. https://doi.org/10.1177/0309132511435352

Lorimer, J., Hodgetts, T., Barua, M. (2017) Animals' Atmospheres. *Progress in Human Geography* 0309132517731254. https://doi.org/10.1177/0309132517731254

Lorimer, J., Srinivasan, K. (2013) Animal Geographies, in: Johnson, N.C., Schein, R.H., Winders, J. (Eds.), *The Wiley-Blackwell Companion to Cultural Geography*. John Wiley & Sons, Ltd, pp. 332–342. https://doi.org/10.1002/9781118384466.ch29

Lulka, D. (2009) The Residual Humanism of Hybridity: Retaining a Sense of the Earth. *Transactions of the Institute of British Geographers* 34, 378–393. https://doi.org/10.1111/j.1475-5661.2009.00346.x

Marvin, G. (2005) Guest Editor's Introduction: Seeing, Looking, Watching, Observing Nonhuman Animals. *Society & Animals* 13, 1–11.

Mason, V., Hope, P.R. (2014) Echoes in the Dark: Technological Encounters with Bats. *Journal of Rural Studies* 33, 107–118. https://doi.org/10.1016/j.jrurstud.2013.03.001

Maurstad, A., Davis, D., Cowles, S. (2013) Co-Being and Intra-Action in Horse–Human Relationships: A Multi-Species Ethnography of Be(com)ing Human and Be(com)ing Horse. *Social Anthropology* 21, 322–335. https://doi.org/10.1111/1469-8676.12029

Meijer, E.R. (2017) *Political Animal Voices*. University of Amsterdam, Amsterdam, The Netherlands.

Morelle, K., Podgórski, T., Prévot, C., Keuling, O., Lehaire, F., Lejeune, P. (2014) Towards Understanding Wild Boar *(Sus scrofa)* Movement: A Synthetic Movement Ecology Approach: A Review of Wild Boar *(Sus scrofa)* Movement Ecology. *Mammal Review* 45, 15–29. https://doi.org/10.1111/mam.12028

Moss, C. (1982) *Portraits in the Wild: Animal Behavior in East Africa*. The University of Chicago Press, Chicago.

Neo, H., Ngiam, J.Z. (2014) Contesting Captive Cetaceans: (Il)legal Spaces and the Nature of Dolphins in Urban Singapore. *Social & Cultural Geography* 15, 235–254. https://doi.org/10.1080/14649365.2014.882974

Ogden, L.A., Hall, B., Tanita, K. (2013) Animals, Plants, People, and Things: A Review of Multispecies Ethnography. *Environment and Society* 4, 5–24. https://doi.org/10.3167/ares.2013.040102

Pile, S. (2010) Emotions and Affect in Recent Human Geography: Emotions and Affect in Recent Human Geography. *Transactions of the Institute of British Geographers* 35, 5–20. https://doi.org/10.1111/j.1475-5661.2009.00368.x

Power, E. (2008) Furry Families: Making a Human–Dog Family Through Home. *Social & Cultural Geography* 9, 535–555. https://doi.org/10.1080/14649360802217790

Rogers, C.R. (1961) *On Becoming a Person: A Therapist's View of Psychotherapy*. Houghton Mifflin, Boston.

Swanson, H.A. (2017) Methods for Multispeciesanthropology: Thinking with Salmon Otoliths and Scales. *Social Analysis* 61, 81–99.

Urbanik, J. (2012) *Placing Animals: An Introduction to the Geography of Human-Animal Relations, Human Geography in the Twenty-First Century. Issues and applications*. Rowman & Littlefield, Lanham.

Van Dooren, T., Kirksey, E., Münster, U. (2016) Multispecies Studies: Cultivating Arts of Attentiveness. *Environmental Humanities* 8, 1–23.

Van Dooren, T., Rose, D.B. (2012) Storied-Places in a Multispecies City. *Humanimalia* 3, 1–27.

Vieira, M. (2011) *Black Bear Data Analysis Unit Management Plan, Northern Front Range Unit DAU B-3*. Colorado Division of Wildlife, CO.

Whatmore, S. (2002) *Hybrid Geographies: Natures Cultures Spaces*. Sage, London.

Wolch, J.R. (1998) Zoöpolis, in: Wolch, J.R., Emel, J. (Eds.), *Animal Geographies: Place, Politics, and Identity in the Nature-Culture Borderlands*. Verso, London; New York, pp. 119–138.

WWF (2006) Factsheet: Human-Animal Conflict.

Part III
Visualising

8 Shared Sensory Signs
Mine Detection Rats and Their Handlers in Cambodia

Darcie DeAngelo

Welcome to Cambodia

The sun glared on the Thai side of the border, more brilliantly where the street is paved than on the Khmer dirt road in Poipet. We peered across it. Some of us, shorter than others, were standing on our toes. We were watching out for Seng,[1] the Khmer man who had managed the rat's importation across the border into Cambodia. The NGO staff and I had our cameras ready. APOPO,[2] an NGO that uses mine detection rats, wanted us to document the rats when they first arrived in Cambodia, but the border was crowded with people, vans and cars. We were looking for a man who would be pulling a cart behind him – a cart like the ones that were used to sell spicy snails on the street. He was a Khmer man Seng hired to do the job of transporting the rats in temporary cages made of jerry cans across the border. By law, animals cannot be imported via an air-conditioned van into Cambodia.

Seng stayed with the rats all night – they spent the night in an air-conditioned room. He later described them with glee to Liz and to Edward, the animal behaviourialist who worked with APOPO in its Tanzanian headquarters. We were all speaking English because Liz and Edward were both anglophones. "I couldn't sleep! They squeaked all night—sometimes fighting with each other through their cages. One bit me through the cage".

"Because they don't sleep at night—they're nocturnal. I think they're also on high alert because of all the different smells and sounds", Edward said. Among all of us, he was the most worried – the rats were so sleepy when they disembarked from the plane. He was concerned that it was too hot for them and that they had been through so much trauma.

Samboth, a Khmer APOPO supervisor who already met the rats in Tanzania, came running to us. "Hey! You missed us", he said. Behind him was a man who looked like a farmer, with his brown clothing hanging off his thin limbs. He had a farmer's hat – a wide brimmed straw hat to keep the sun off his face. And behind him was the cart.

We all rushed to see the rats. We peeked under the canvas that covers the cages, but Edward shooed us to the van where we could cool them off. The three men began transporting them one by one. The rats were stirring, brown creatures in bright yellow jerry cans that were cut to make a box. Red wire

DOI: 10.4324/9781351018623-11

covered the jerry can's entrance so they couldn't jump out. They stuck their noses through the red wires, sniffing. Their little hand-like paws clutched at the wires. "Welcome to your new home", Samboth said, "very different from Tanzania. Not the same place".

This chapter draws from the multispecies ethnography above to better understand relationships among mine detection rats and their human handlers in Cambodia. How can ethnography show the rats' point of view? From the first day the rats arrived in Cambodia in April of 2015, I conducted participant observation, videography and audio recordings of how the handlers and rats learned to negotiate the mine detection objectives with each other; how the rats learned to navigate a new climate and work environment; and how the foreign supervisors trained the Cambodian demining team.

This chapter addresses sensory ethnography methods, that is, visual, tactile, olfactory, and auditory representational modes through descriptive text, video and audio, in order to show a multispecies perspective in ethnography. The primary objective of such methods is to attend to that which is beyond words, with the understanding that animals and humans communicate semiotically beyond symbolic language. Using multispecies ethnography through video, audio and sensory representations, I was able to capture data that escapes the limits of words, thus addressing communication that is "more than human". My videos show multiple and contrasting perspectives in the minefield. For, while the Cambodian handlers called the rats "little sisters" and "best friends", the rats struggled against their harnesses and portrayed fear and exhaustion during mine detection.

Shared Sensorial

Like most anthropologists, I base much of my analysis on human behaviours and human words, but these methods betray me when it comes to nonhuman interlocutors in my field site. During research among a platoon of Cambodian landmine detectors, I conducted interviews and participant observation with humans who described to me their fears about the government, the minefield and also their love for the newly imported landmine detection rats. Based on the ways in which the humans related to the mine detection rats, the bonds between human and nonhuman seemed certain: love. Deminers told me that the rats were their "very special friends", one deminer said that she felt that her rat was her "little sister". What of the rats and their perspectives? I attempt to trouble this notion of love between humans and nonhumans through sensory ethnography.

Sensory ethnography allows me to get at both human and nonhuman perspectives on these relations and to explore uncertainties. Based on Peircean semiotics, signs are not limited to representational modes defined only by other signs, that is, language. Representational modes beyond words are characterised by their connections to the objects they represent, a characterisation which allows nonhuman beings who share the experiences of those objects the possibility of sharing those representations. Many anthropologists have argued that we must move away from a purely linguistic analysis of so-called "culture". Instead, researchers should analyse and incorporate other kinds of signs that are not linguistic, signs made from media like images, poetry, sounds, fiction, video and film. Symbolic reference alone cannot capture nonhuman (or human) experience. In cases where animals are primary interlocutors, anthropologists must find different media to understand signs (Kohn 2013, p. 40).

The primary objective of such methods is to attend to that which is beyond words, with the understanding that animals and humans communicate semiotically beyond symbolic language. By this I am referencing an understanding of semiotics as outlined by Charles S. Peirce because, while humans use words in the Saussurean sense, as arbitrary symbols in a complex system of representation (de Saussure 1959), animals and humans share modes of representation that are beyond (or before) the mediation of words. Different kinds of words, or rather, different kinds of signs, as in those visceral signs that directly reference reality (e.g., images, dance, video, sounds or poetry), which do not follow the system of words as symbols, are called icons or indexes (Peirce 1998, pp. 4–10; see Concept Box 8.1). From the visual and visceral methods of video and audio recordings, I was able to follow the learning curve of the deminers handling the rats that also spoke to a gradual progression of love and comfort between human and animal, rather than an immediate love at first sight the deminers portrayed.

Peirce outlines the modes of representation between interpretants and objects by differentiating a tripartite system of signs. Each sub-system differs by how interpretants make sense of a sign's meaning, that is, how they draw the connection between a sign and its object (Peirce 1998, p. 271). Different types of signs function differently than the one-to-one definitions of symbols and these are called icons and indexes (1998, pp. 5–7). People interpret symbols by defining them in relation to other symbols. For example, the word "cat" is a symbol of a particular kind of animal; however, the content of the word "cat" is defined by how the interpretant relates "cat" to other symbols like "dog", "kitty", "pet", etc. "Cat'" has no inherent relationship to the milk-loving domestic animal it refers to; it only derives this meaning by virtue of its connection with other inter-related symbols.

Icons and indexes, on the other hand, operate in a very different manner, as both are signs that are connected to that which they represent. An index works through a causal relationship, sometimes in the form of a sensory input that indicates what it represents; for example, a column of smoke signifies a fire to an interpretant with an index. Interpretants understand icons with the simplest representational mode; icons look, sound or in some way resemble what they signify. Icons signify based on shared qualities of the world, or rather, with the objects that they represent, even when that object does not exist (Peirce 1998, p. 6). Symbols are the only signs that are unique to humans because life, both biological and spiritual, is semiotic (Kohn 2013, p. 8). According to this logic, humans can share icons and indexes with other semiotic beings such as animals. Animals are semiotic beings who can understand modes of representation beyond words. As an example, the philosopher, Vinciane Despret (2004) describes experiments in the 1960s which portrayed the unseen communication of lab rats and human experimenters. When a human experimenter expected that a rat was from a "bright" group, the rat solved the maze puzzles better than the rats labelled as part of a "dull" group. She reframes these experiments not only as human labels and unconscious expectations that alter a rat's performance, but also as ways in which the rat and human respond to each other's nonverbal representations, which are categorised semiotically as iconic or indexical. Kohn's example of an icon is a scarecrow, which signifies a raptor predator in the Amazon to parakeets and to humans. The important note here is that the raptor icon does not approximate a "realistic" version of a raptor from a human point of view, but rather, a version of a raptor from a parakeet's point of view (2013, p. 115). With this icon, humans can "know something of what it is like to be a parakeet" (116), by creatively guessing at what a parakeet's interpretation of the raptor figure would be. Since the parakeet is a self, it can understand icons; the raptor scarecrow is a product of a shared semiosis with the living selves of the forest. What this comes down to is the fact that humans and rats can share sensations of the world, allowing for an ethnography that incorporates such shared sensoria.

Concept Box 8.1: Signs

Signs in semiotics are understood as *modes of representation*, that is ways in which worlds and existences are portrayed (by humans and non-humans). Ferdinand de Saussure developed signs as integral to the formation of language, but in his theory, he separated modes of representation from the material world. For Saussure, signs exist only because of humans and they are limited to human minds, thus the central relationship for him is between the human and the sign, which does not have any linkage to the material world. In contrast to the Saussurean relationship between subject (person) and object (sign), Charles S. Peirce establishes a more complicated relationship of a sign, that is, a sign, an object, and an interpretant. Peirce calls the mode of representation, the "sign" a subject of the sign (so symbols are also kinds of subjects) (Peirce 1998, p. 477). This means that the sign and the interpretant interact to produce the object. In the subject–object dichotomy established by Saussure, only interpretants would be considered subjects. Peirce proves useful when theorising nonhuman modes of representation because he allows for non-linguistic signs, which is why he has been taken up by biosemiotic theorists. The neuroanthropologist, Terence W. Deacon, suggests in his seminal text, *The Symbolic Species: The Co-evolution of Language and the Brain*, that the development of human language parallels other nonhuman modes of representation (1997).

My sensory ethnography constitutes a shadow of the love the human dem-iners expressed for their mine detection rats. Ethnographic videos and audio complicate the love that the deminers professed. The video is also one way to approximate a rat's perspective because it deals with sensory experiences of the ground, the grass and the size perspectives of a small rodent. Another way I show these uncertainties about love is by interspersing images of human-animal relations in the video and in my writing. Most of these images, though not all, focus on the rats. I include these stories so that I can offer a different perspective from humans, problematising the anthropocentrism of human portrayals. Some of these stories are experimental, texts that imagine what it is like to be a rat, taking my cues from the biological sciences and behavioural sciences as they pertain to rats. For example, studies show rats do not pass the mirror test,[3] rats can sense infrared spectra, rats have incredible senses of smell, and rats cannot see very well (Hanson 2004). With this sen-sory ethnography, I presume that a human, like you and me, can imagine what it is like to be an animal (Coetzee 2016, pp. 31–33). This also presumes that both nonhuman animals and human animals can think, and in that they share certain semiotic modes of representation (Kohn 2013). With these semiotic modes of icons, I attempt to disrupt the straightforward grammar of my text to better explore relations between nonhumans and humans. That said, I do not suppose that the imaginings I present are entirely correct, nor do I claim that such things can be captured entirely by my ethnography. Instead, I only seek to offer another way of showing human-animal relations in a minefield.

Unspoken Struggles

After the rats arrived, the handlers were hired and lodged in bedrooms that shared a wall with the rat kennel in the Roluos district. This kennel was a government building next to another, bigger government building that served as a meeting place for mine action conferences held by the Cambodian state. In that building, landmines and other bombs were displayed behind glass in a huge conference room. The deminers slept on cots in a tiny neighbouring concrete house. The rats altered the deminers' sleeping schedules – both rat and human must begin work before dawn since it is far too warm for the rats in the afternoon. Each morning the deminers smeared sunblock on the sensi-tive, translucent ears and the hairless paws of the rats. The rats were new actors in the ecology of the minefield – they were imported from Tanzania. They were Giant Gambian Pouched Rats. The Cambodians had never met rats like these. "Before I met these rats, I thought of rats as pests", said one deminer in an interview, "but once I met them they are my very special friends".

The rats in Cambodia were usually small rats, urban rodents that could be found all over Southeast Asia. The landmine detection rats were different from these common street rats – they were of the phylogenic order *rodentia*, the genus, *cricetomys*, and known by their common name of either "African giant pouched rat" or "Gambian pouched rat". They weighed, on average, 1–2 kilos (2–5 lbs.) and they could grow up to 0.9 meters (3 feet) long, although their two-toned tails made up half that length. They looked like typical urban rats except bigger – they had everything you would expect a rat to have: circular folded translucent ears, beady black eyes, tiny hand-like paws and pointy noses that shivered constantly since scent was their most dominant sense. Because APOPO's rats were ground rats, their sense of smell had evolutionarily developed to be more sensitive than arboreal rats. As rodents, they were both smart and trainable, but because they were giant rats, they were large enough to be handled. Unlike other rats, such as the commonly seen Norwegian rat, Gambian pouched rats had big cheeks like squirrels. The rats sometimes stuff these cheeks full of food for safekeeping, giving them fat faces.

Landmine detection with rats began with cleared pathways between minefield pits of 200 square meters. The rat was attached to a harness which guided it along a string between two deminers who stood in the cleared pathways across from each other. When the rat smelled TNT beneath the soil, she scratched twice and then the deminers marked that point with a flag. Later, a detonation team exploded the ordinance safely. The work was painstaking. The deminers mapped the minefield on paper and marked it with targets, meanwhile walking step by step alongside. The rats scurried between the humans but sometimes stopped on their own to groom themselves, which led the humans to make jokes about the rat not needing to stop work in the middle of the pit to "take a shower" because she was already so beautiful.

If I had limited my ethnography to the testimonies and behaviour of humans on the minefield, I would have had a straightforward story of love between humans and nonhumans. The ethnographic filming reveals more of the rats' perspectives. For example, in the minefield, the rat, Victoria, struggled with her harness. The handlers tried but failed to get the pink straps on both her forelegs and she squirmed and pawed at them. Her big teeth gnashed, but she was silent (to the humans) because rodents mostly communicate with squeaks and chirps that humans cannot hear. And the Cambodians did not say anything either, but gritted their teeth as they tried to clip the plastic. They called Hien, the Khmer deminer, to help them.

Hien took Victoria, who was bigger than both his hands, and squatted over her. His shadow surrounded her. She leapt out of his hold and paused to look where to go, but we were all crouching in a circle with her in the middle: two of the handlers, myself, and Hien. Her nose twitched and her whiskers vibrated constantly. She began to wash herself, licking her forepaws and rubbing them against her ears.

"She is afraid", Hien murmured, and he reached for her again. She ran as and he slammed both palms on the ground and, rather suddenly, shouted, "Calm down!" This didn't work.

"Well, don't shout at her", I said but automatically regretted it because I think it must have sounded rude.

Figure 8.1 Photo by Darcie DeAngelo. Hien, the deminer, taking Victoria to place
   her in a transport cage.

Victoria ran even faster in the other direction and Hien stood up and
scooped her up by her midsection and placed her in her transport cage (see
Figure 8.1). He placed the barred cover on top and walked to where Omary,
the short Tanzanian consultant, was holding one side of a measuring tape
across from his trainee, Sarim.

> "Omary", Hien said, placing Victoria's clear portable cage on the ground
> in the shaded cleared area outside of the mine-contaminated borders.
> "This is for you".
> "Why not you?" Omary asked without looking at him, watching the
> rat, Janthina, while she scurried between him and Sarim.
> "This is for you", Hien repeated.
> "I know, but, why not you?"
> "This is for you, why don't you understand?" Hien's voice started to
> rise. "You are the manager; you are in charge. This is your job!"

Then Omary relented and looked at him and said, "Okay". Hien walked off
to the buildings and another Khmer supervisor followed. Omary and Hien
had been having issues with each other. Meanwhile, Victoria balled her body
up in the cage and closed her eyes.

The above dialogue shows some interesting tension between the Khmer
deminer, Hien, and the Mozambiquan consultant, Omary, apparent through
the written record of their words. To understand Victoria, however, we need
methods of a (multi-)sensory ethnography. It is the image of Victoria running from Hien, the sight of her balled up in her cage, that more clearly

demonstrates how the animal understands the experience. She trembles and, in the image, she is in the process of fleeing from Hien's hands. The hand grasps her, and it is so large compared to the rat's body. We can imagine how a human must seem huge to Victoria. The camera, like the humans, looks down, providing insight into how it must feel as though humans loom over her. She is scared. That is clear from the use of sensory ethnography.

Unheard Jokes

Video served as a method to capture moments from a distance, as well as intimate interactions between humans. The process of training the rats led to these subtle interactions between humans related to how humans objectified the rats and each other. Organisation scholars Hamilton and McCabe (2016) show how chickens become "de-animalized" (332) in order to be processed as protein sources in factories. This process de-emphasises the animal's subjecthood to instead frame the animal as an object and a commodity. One day at the minefield, the rats were being taken one by one from the minefield to the kennels. The groundskeeper, a fellow governmental employee of the deminers, offered to help. When the deminer handed the plastic container holding a rat to the groundskeeper, he did so very carefully, so carefully that the groundskeeper mocked his care by pretending to drop the plastic container. That is, he dropped it a little and caught it before it hit the ground. The deminers laughed, but the camera caught this moment. The rat inside the plastic container scrambled and suddenly became very still.

This moment was not a good moment for the rat, nor perhaps the deminer and groundskeeper. They were out of sight from their supervisors but realised after the joke that I had been shooting the video camera. I remember catching the eye of the man who play-dropped the rat, but he quickly looked away.

The following week, one of the rats died.

The day of his death was a storm of stress. Chann called me before dawn to tell me that I would be picked up later. He did not say why or where, and then, Liz (the animal behaviourist) called me to meet her mid-morning in the city so that we could drive over to the minefield together. I met her at the government building in the city, walking down a muddy river path to where she was smoking and frowning outside. She shook her head when she saw me.

> "One of the rats is at the vet—we have to go". She pointed to the car and then I followed to the passenger side. She didn't look at me.
> "Which rat?" I asked.
> "Håvard".
> "Oh! He was biting a lot last week—maybe he's sick?"
> She didn't say anything as she navigated the jeep along unpaved narrow roads to the river where there was a high-end shopping area. As we drove along the river, she said, almost to herself, "I bet he's already dead. I bet they're just not telling me".

We arrived at the city's only pet store – a small boutique that catered to foreigners. It had windows decorated with stickers of skinny cartoon women holding leashed poodles. The door was open. Past aisles of plastic cat toys and brightly coloured bags of brand name pet food, the Khmer vet, Hak, stood at the back of the store looking down into one of the rat transit cages on a filing cabinet. It was transparent, lined with paper towels and the size of a shoe box. Seng, Liz's assistant, and Hien were there. Chann was still at the minefield.

As we walked in, we could see Håvard on his back in the cage, mouth partly open and legs curled tight to his white belly. Hak had tears in his eyes. As soon as she saw the rat, Liz left the boutique to call Tanzania. Hak told me in Khmer that Håvard had died in the car. He mimicked Håvard's last noise, a gulping gurgling cry. The woman in the store looked sympathetic but explained that there was no vet here who could do an autopsy.

We all drove back to the minefield with Håvard's corpse.

People were anxious after Håvard died. The foreign staff muttered that the Cambodians were rough with animals and I was nervous, too. I told Liz about what I had seen in the minefield during training. I wasn't sure which rat had been joked with, but it if was Håvard, it seemed to me as if it's something I should say – it seems like it may be one of the possible reasons he died so mysteriously. I also thought that they must act more carefully with the rats. I felt worried about the animals.

We had another meeting the next day. A lot of the meetings were like this: we sat in the kitchen and the supervisors critiqued the staff. I found that my mind wandered – we were under the shade of a barn-like structure with two walls. The kitchen was behind a little locked wooden door with rotting planks, and a big water tank overlooked a little stream where chickens strutted. While the supervisors pointed at people and yelled at them in front of everybody, I watched the movement of grass in the wind or listened to the tittering bird that was something like a grouse, swooping down to the field in spirals to find prey. Chann smiled and pointed at me in the midst of chiding his staff for not paying enough attention to the rats. "Darcie has some things to say".

I was surprised at the weight of the gazes on me. I didn't know what he wanted me to say but I spoke as positively as possible. "I think it's so wonderful that some of you feel close to your rats. Some of you say they are your friend or your little sister".

"Yes, but, Darcie", Chann said, "didn't you see something?"

I realised then that he was trying to get me to join in on the chiding. I felt embarrassed that the supervisors knew what I had said to Liz. I said, "Oh! Okay, yes. I think sometimes it's easy for humans forget that the rats are small. What seems like a small distance to us is actually very big for them".

Chann seemed a bit confused and I felt as though I had not been as critical as I should have been. Chann, still with a stiff jaw as if I was not playing the right game, repeated what I said in Khmer, and added that I saw one of the staff who, was not a rat handler, handled one of the rats and pretended to drop it. Were the deminers angry with me? While they smiled at me after I said what I had to say, that did not mean they were not angry.

I include this moment of fieldwork to briefly consider the ethical dilemmas in my fieldwork with human-animal studies. In the end, the footage I captured showed a different rat (not Håvard) being play-dropped. So the joke in this case could have gone unheard and unnoticed. The filming brought up these ethical issues in ways that only happened through its media – of course I could have told Liz without the recording what I had seen, but it was the technical capabilities of the lens that allowed me to see the joke in a way that I would not have noticed.

I do not yet have an answer for what would have been the most ethical choice in this moment – to keep the footage, which I had full research ethics board approval and interlocutor consent for; to seek renewed consent from the subjects given that they had done something which may have embarrassed them; or to express my concern for the danger the rats were facing.

The joke would have been unheard, however, if I had not had the footage. The groundskeeper was silent on the screen as he dropped the rat cage, forcing the rat to scramble in her cage. The deminer looked nervous. The video allows for a re-interpretation of the straightforward love the deminers profess for the rats and speaks to the process of de-animalisation that often occurs when humans work closely with animals. It also points to the taboo moments that do not get spoken aloud.

Love

And yet, the deminers did learn to love the rats. This is clear from the ways in which they began to cuddle them, the ways in which they spoke gentle words of affection for them, and how they professed to love them. What of the rats? Do the rats love the deminers? The rats cannot tell us how they feel. They do not rely on words for communication. I propose we share enough sensoria with rats to understand what it is like to be a rat. Can we then also say that we know what they feel?

The premise of the question can already be undermined since feelings of love are often thought to be feelings exceptional to humans. Putting aside questions of emotional capabilities that may (as neurologists suggest (Panksepp 2005)) or may not exist in shared neurological structures, we can revise the question to simply ask if the rats reciprocate whatever connection the humans feel for the rats. In the ethnographic footage, I watch the rats struggle in their harnesses, flee from the humans who sought to train them, and lumber with the weight of dehydration and fatigue. As I watch the footage and the months of training, though, the rats clearly learned to be comfortable in this new setting. Was this comfort, love?

When a rat finished detection work in their practice pits, the handler unclipped their harness and guided them to their transport cage. The handler patted her legs and the rat bounded after her – an action that delighted anyone watching, though it took some time for the rats to follow the Cambodian staff as opposed to the more experienced Tanzanians. Issac, a huge male, was the exception, because he had always followed his handler, Moch.

Sometimes I asked questions with my camera, but it always began so early, just 5:30am, and, along with everyone else, I was very tired and focused on the rats. I was mostly silent at the minefield. The workers were silent too, as they watched the rats wander between them on twine and measured each step along the pits. The pits were set up in long rectangles with flags and measuring tape strung between posts. The tape marked the border of where the landmines are – outside the border we can walk freely. Even though these mines were deactivated, we were meant to be as careful on the minefield as though the danger was omnipresent. We avoided the pits so that it became a habit. On each side of the pit, one deminer stood across from the other, with a rope looped around one of their ankles. The rat, wearing a little harness that it occasionally stopped to struggle with, ran along the rope and each deminer held one end of another measuring tape, to keep track of where in the pit the rat has found a landmine.

The rats were dainty as they hurried across each pit's lumpy soil while the deminers looked sluggish. They all stomped in their military boots except for Moch, who at the time wore rubber galoshes (the government had not yet given her entire uniform) which made her cautious on the slippery grass.

I gravitated to her one day and I noticed that Moch adored her rat, Issac. Issac lumbered, he was not dainty like the rest of them. He weighed 2 kilos and his hind legs looked to me as if they spread with flapping skin.

Issac scratched at every mine in the practice pit. "He is a good rat, very proper and very smart", Moch told me. She unlatched his harness and walked ahead of him, looking down. "Come here, little sister, come here, little sister".

He followed the sound of her voice. Issac, like all the rats, was mostly blind but he could hear and smell very well. He jumped over the grass to follow Moch in her rubber boots. I crouched with Moch as she helped Issac into his labelled transport cage. On the footage and in photographs, Issac nibbled Moch's neck and cuddled with her. "I just feel as though he is my little sister", she said, "even though I know he is male".

She did not mention reincarnation, but perhaps this was what she meant. In the frame of sensory ethnography, what was important is that Moch and Issac share the sense of touch and its shared sensoria is apparent in the visuals. Moch and Issac clearly showed affection for each other (Figure 8.2).

Multispecies Sensory Ethnography

While the sensory ethnographer Castaing-Taylor uses the word "human" when he says that we "express ourselves through images as well as through language" (1996, p. 85), in his recent films, the imagistic indexes were not peculiar to the humans portrayed. In both Sweetgrass (2009) with Ilisa Barbash and Leviathan (2012) with Verena Paravel, the filmmakers focused on the experiences of animal and humans. The filmic images, both sonic and visual, communicated human and nonhuman entanglements by following sheep herders and sheep in Sweetgrass (2009) and fishermen and Atlantic Ocean organisms in Leviathan (2012).

Figure 8.2 Photo by Darcie DeAngelo, Still frame of a landmine detector and their rat, 2015.

Alanna Thain (2015) describes how the filmmakers of *Leviathan* and *Sweetgrass* give nonhumans perspectives in their films by using iconic and indexical signs that nonhumans present. For example, the opening scene in Leviathan (2012) shows a black screen with light bouncing off a long rope that lunges in circuitous diagonals across the screen and makes sounds that come from machinery but are so loud and monstrous that they seem as though they come from a nightmarish creature, allowing even the technologies to be included as perspectives. Other shots come from a series of digital cameras attached to pulleys on the masts and sides of a commercial fishing boat. These shots engage, for the most part, with nonhuman experiences in the ocean, such as underwater sea creatures like starfish and cod as well as birds like seagulls, as they live and die. One particularly long shot focuses on the struggles of a bird to find refuge on the boat as it is pounded by a heavy storm, which eventually pushes it off the boat's deck into the churning ocean, presumably to its death. Even as fish are slaughtered by humans, much of the shots of slaughter are from the fish's points of view with the camera literally in puddles of water where fish slide on the deck, suffocating.

Similarly, sensory ethnographers use sonic images to represent nonhumans as semiotic selves. Ernst Karel, the sound engineer for the film Sweetgrass (2009), explains that the film's sound has very intentionally been mixed so that the human cowboy's voice is not prioritised as "a privileged position" over the sheep (Nayman 2013). The bleating in the film can sometimes overpower the voices of the few human characters who are herding hundreds of them to be slaughtered from Colorado to a neighbouring state, but it clearly expresses sheep distress, sheep communicating with another, and the sensory experience of being there with the sheep and the cowboys.

These films use images as iconic and indexical composites which allow the ethnographers to better communicate and interpret things beyond words: the fear and rocky violence of a commercial fishing boat, the solitude of being a lone cowboy amidst a bunch of sheep, and perhaps even the bestial fear of sheep as they are attacked by wolves one night. From the films emerges a semiotic consideration for the animals as well as the humans that communicates the experiences of both, experiences that exist before or beyond the mediation of words. Therefore, while Castaing-Taylor asserts that images are not the only, nor necessarily, better way to communicate human experience (Taylor 1996, p. 85), it would seem, through the films, that sensory ethnography represents human-animal experiences in a unique way because animals do not use symbolic systems, but do share with us icons and indexes as modes of representation (Peirce 1998, p. 280; Deacon 2012, p. 442; Hui et al. 2008, p. 87; Kohn 2013).

In my footage of mine detection rats and their handlers, I attempt a similar sensory multispecies ethnography. Through footage of a rat who struggles, eats, sleeps, and cuddles, we get at unique perspectives of both the rat and the human interactions. It becomes possible for us to understand the rat's fear, hunger, fatigue and (possible) love when we utilise sensory ethnographic methods to see things from the rat's eye view. With the camera and audio recording we can better answer the question: what of the rat? With sensory modes, we, as ethnographers, can better represent nonhuman perspectives of our shared, entangled worlds.

Notes

1 All names and identities have been changed and disguised according to ethical guidelines for the protection of informants.
2 The acronym for Anti-Persoonsmijnen Ontmijnende Product Ontwikkeling (in English, Anti-Personnel Landmines Mine Removal Product Development), the NGO that uses mine detection rats.
3 The mirror test is given to determine whether or not an animal or human can recognise itself in the mirror. It is said to indicate a theory of mind. The test involves marking an animal's face and presenting it with a mirror. If the animal sees the mark and attempts to wipe it off, it is said to give some evidence that the animal has a sense of self. However, the results of these tests have been deemed questionable by some researchers, especially when they involve animals that do not use their sense of sight as much, as with dogs or rats (Bekoff & Sherman 2004). And humans have often misinterpreted the results of the mirror test, not noticing that some animals, like gorillas, are so embarrassed by the spot on their face that they will wait until everyone leaves before they wipe it off (Koerth-Baker 2010).

References

Barbash, I. and Castaing-Taylor, L. (2009) *Sweetgrass*.
Bekoff, M. and Sherman, P.W. (2004) Reflections on animal selves. *Trends in Ecology & Evolution*, *19*(4): 176–180.
Castaing-Taylor, L. and Paravel, V. (2012) *Leviathan*.

Coetzee, J.M. (2016) *The Lives of Animals: The Lives of Animals [Princeton Classics]*. Princeton University Press.

Deacon, T.W. (1997) *The Symbolic Species: The Co-evolution of Language and the Brain*. WW Norton & Company.

Deacon, T.W. (2012) *Incomplete Nature: How Mind Emerged from Matter*. WW Norton & Company.

Despret, V. (2004) The body we care for: Figures of anthropo-zoo-genesis. *Body & Society*, *10*(2–3): 111–134.

Hamilton, L. and McCabe, D. (2016) 'It's just a job': Understanding emotion work, de-animalization and the compartmentalization of organized animal slaughter. *Organization*, *23*(3): 330–350.

Hanson, A. (2004) "The rat's sensory World." Articles and research. *Rat Behavior and Biology (Anne's Rat Page)*. http://www.ratbehavior.org/RatSensoryWorldMain.htm

Hui, J., Cashman, T. and Deacon, T. (2008) "Bateson's Method: Double Description. What is It? How Does It Work? What Do We Learn?" In *A Legacy for Living Systems*, pp. 77–92. Springer Netherlands.

Koerth-Baker, M. (2010) Kids (and animals) who fail classic mirror tests may still have a sense of self. *Scientific American*. https://www.scientificamerican.com/article/kids-and-animals-who-fail-classic-mirror/#; accessed 3/4/2022

Kohn, E. (2013) *How Forests Think Toward an Anthropology Beyond the Human*. University of California Press.

Nayman, A. (2013) RIDM Dispatch #3: At the heart of the Sensory Ethnography Lab. http://povmagazine.com/blog/view/ridm-dispatch-3; accessed 11/24/14

Panksepp, J. (2005) Affective consciousness: Core emotional feelings in animals and humans. *Consciousness and Cognition*, *14*(1): 30–80.

Peirce, C.S. (1998) *The Essential Peirce: Selected Philosophical Writings* Volume 2 (1893–1913) edited by the Peirce Edition Project. Indiana University Press.

de Saussure, F. (1959). *Course in general linguistics*. Philosophical Library, New York.

Taylor, L. (1996) Iconophobia. *Transition*, *69*: 64–88.

Thain, A. (2015) A bird's-eye view of L leviathan. *Visual Anthropology Review*, *31*(1): 41–48.

9 Doing multispecies ethnography with mobile video

Exploring human-animal contact zones

Katrina M. Brown

Introduction

"If we appreciate the foolishness of human exceptionalism then we know that becoming is always becoming *with*, in a contact zone where the outcome, where who is in the world, is at stake" (Haraway, 2008, p. 244).

In the burgeoning field of human-animal studies, significant importance has been ascribed to the multispecies *encounter*; spaces in which the differently embodied, differently lingual, and differently encultured and empowered meet, mingle and mutually constitute each other (e.g. Kohn, 2007; Despret, 2004). Haraway (2008) and others (e.g. Sundberg, 2006) consider such sites as "contact zones", foregrounding the contact, mutual affect and co-becoming through which human and nonhuman animals entangle and are made available and vulnerable to each other (see Concept Box 9.1). They all underline the consequences particular contact zones have regarding the flourishing of, or harm to, individuals, species and wider ecologies. This makes the how and why of their intricate interspecies relationality both vital and urgent for our greater understanding. Work done in multispecies ethnography makes the case that in order to craft ethical, shared lives with nonhuman others, we need to learn to notice, listen to and pay closer attention to animals (Kirksey & Helmreich, 2010; Ogden et al., 2013; van Dooren et al., 2016;) and indeed find better ways to let animals themselves "speak" more on their own terms (Bastian, 2017; Bastian et al., 2017; Despret, 2016).

The call for creative more-than-human methodologies is therefore becoming ever louder (Hamilton & Taylor, 2017; Dowling et al., 2017). There is a need to allow the articulation and noticing of animal ways of being and knowing that often entail: (a) spatiotemporal patterns of gesture, movement and comportment that can be subtle, sophisticated and often fleeting and unpredictable (Lorimer, 2010; Brown & Dilley, 2012) as well as part of complex mobilities and their politics (Hodgetts & Lorimer, 2018); (b) somatic expression of intelligence and language using ways of articulating and responding that are viscerally and multi-sensorially registered, often using sensory hierarchies humans would find unfamiliar (Grandin, 2006; Kohn, 2013); (c) attunement to animal atmospheres (Lorimer et al., 2019) and intricate affective dynamics (Despret, 2004); and (d) locatedness within complex ecologies

DOI: 10.4324/9781351018623-12

Concept Box 9.1: Contact Zones

The concept of a contact zone was introduced by Pratt (1991, 1992) to denote a space of cross-cultural encounter between different people who "meet, clash and grapple with each other, often in contexts of highly asymmetrical relations of power" (Pratt 1991, p.33). She identified in these junctures both vulnerability and generativity in the exchange, as despite the possibility of actual or threatened violence, the onus was on subjects to develop ways of communicating across different languages and ways of knowing. A key feature of a "contact" perspective is emphasising the possibility of mutual transformation, "how subjects are constituted in and by their relations to each other" (Pratt 1992, p. 7). Rather than living out simplified or predetermined subjectivities of dominance and capitulation, each culture can select elements and create something new in a process she calls *transculturation*. Crucially, therefore, conceptual space is made for the agency of those typically considered to be less powerful and for attention directed to the role they play in complex, fluid co-becomings at thresholds of difference.

Haraway (2008) extends Pratt's idea of contact zones to encompass sites of interspecies contact and fleshy encounter in which animals and humans bring each other into being – and even become more skilled together – across different power gradients and ways of knowing. She highlights contact zones as key sites where sensory and affective relations develop, urging us to attend to the bodily entanglements through which species are made available (or otherwise) to each other, how they articulate and respond to one another (after Despret, 2004). Infusing the concept further with the idea of an ecotone and its edge effects, she foregrounds contact zones as sites of richness, diversity and hybridity due to the interdigitating edges, with the potential to trouble established human-animal and nature-culture binaries.

The key features of contact zones emphasised in multispecies scholarship thus include:

*   Locatedness in an encounter
*   Intermingling at a threshold of difference
*   Uneven relations of power
*   A transformative space in which different subjects are interdependent and co-constituted
*   Difference as potentially generative rather than necessarily or wholly subjugating or causing inertness and passivity
*   Nonhuman agency is taken seriously
*   Attention to bodily, sensory and affective relations through which subjects articulate and respond to one another
*   Consequential as subjects are at stake in terms of harm and/or enrichment
*   Edge effects with the potential of diversity and hybridisation

(Hodgetts & Lorimer, 2015). Dealing with more-than-human ways of knowing is hard enough in environments where humans more tightly try to bound animal practices, such as houses (Power, 2008) and laboratories (Greenhough & Roe, 2011). How are we to "stay with the trouble" (Haraway, 2016) when the trouble often involves complex, dynamic, spatially extensive ways of articulating, knowing and mutually affecting? How are we to give "voice" to more-than-verbal animals when words remain our primary mode of intellectual enquiry? Despret (2016) impresses upon us how our will and ability to ask the right questions *in the best possible way* is crucial for deepening our understanding of nonhuman creatures.

A growing number of scholars have looked to video technologies and techniques as one way in which we might experiment with our grasp of more-than-human relations. Video has long been used by ethologists to study animals in a quest to generate objective accounts of their behaviour (Wratten, 1994),[1] but we are only in the early stages of developing video techniques to understand human-animal encounters reflexively and relationally (e.g. Bear et al., 2017; Collard, 2016; Fijn, 2007, 2012; Goode, 2006; Grimshaw, 2011; Konecki, 2008; Lorimer, 2010). Rather than trying to uncover a pre-existing truth about human-animal interactions, this research foregrounds the ability of moving images to witness, evoke and make oneself more available to the shifting bundle of fleshy, multi-sensory, more-than-verbal materialities, vitalities and affective relations through which cross-creaturely entanglements are made and remade. Kohn (2013) underlines how creaturely communication, including the exchange of meaning across species boundaries, is often inextricably embodied and materially entangled (see Chapter 8 by DeAngelo in this book).[2] Likewise, registering and expressing body language is central to the medium and agency of motion pictures, epitomised by the oft-stated imperative of filmmakers to "show don't tell". Therefore, video methods are full of possibility when it comes to engaging with the variety of ways in which animals "speak" through their bodies by registering motion, pattern and fleshy, muscular form as a series of images with synchronous audio.

With the growing accessibility of Wearable Camera Technology (WCT) – in terms of price, weight, size, and user-friendliness – we now have the opportunity to think more expansively about the possibilities for enrolling video in human-animal encounters, in terms of where the camera can be mounted, where it can travel, and the shooting and audiencing techniques that can be used (Brown & Lackova, 2020; Seegert, 2016) and which can be considered to offer an opportunity for radical experimentation with de-centring human perspectives and new forms of more-than-human "listening" (e.g. Landesman, 2015; Stevenson & Kohn, 2015).[3]

WCT has been shown to have particular merit in engaging the mobile, non-verbal, haptic, visceral and affective dimensions of encounters (e.g. Brown & Spinney, 2010; Duru, 2018; Paterson & Glass, 2018; Sumartojo & Pink, 2017) and is ripe for further exploration as a key mode of multispecies ethnographic immersion. Brown & Lackova (2020) identify three key capacities of WCT that enable its signature generation of dynamic, "first-hand" audiovisual impressions, and its promise for understanding human-animal relations: its ease of becoming securely conjoined with a particular body (part) and/or object; its ability to stay with that body-object whilst in motion, despite the numerous frictions of space, time and the environment; and related forms of promiscuity that can readily entangle with the promiscuous, lively multiplicities and complexities identified as central to more-than-human becomings (after Haraway, 2008; see also Jones, Chapter 4, in this book).

Thus, we ask here: what role can wearable cameras play in making available more-than-human ways of being, knowing and articulating? In addressing this question, it is vital to be mindful of the power relations at work in the generation, circulation and audiencing of visualities (Haraway, 1991; Rose, 2015) and consider critically how we are making animals visible or invisible, and the related implications. Wearable cameras have developed primarily through visual cultures of adventure and extreme sport (Vannini & Stewart, 2017) and surveillance (Adams & Mastracci, 2017; Stein, 2017), both of which invite very particular, non-innocent – often privileging able-bodied, masculine – ways of constituting subjects and objects. The adoption of WCT into wildlife film as "critter-cam" has been critiqued by Haraway (2008) as yet another kind of patriarchal apparatus that works to colonise animal lives and bodies. Yet early ethnographic forays also suggest miniature and wearable cameras herald a new way of de-centring human perspectives; whether through the enchanting YouTube clips of pet cam (Seegert, 2016) or the disturbing, visceral, terrible beauty of Leviathan (Stevenson & Kohn, 2015).

In this chapter, I reflect on the experience myself and colleagues have had with employing mobile video ethnography (MVE) with WCT in examining more-than-human becomings.[4] I discuss how moving images and sound mobilised using Point of View (POV) techniques in particular can help us to witness, evoke and make sense of cross-species entanglements, and indeed "stay with the trouble" (Haraway, 2016) as human and nonhuman animals are conjointly and bodily brought into being in specific contact zones. In particular, I consider the more-than-human ways of "speaking" invited and made available to wearable cameras in our MVE of dogwalking practices in the UK pinewood habitats of the capercaillie (*Tetrao Urogallus*), an IUCN Red Listed ground-nesting bird known to be sensitive to recreational disturbance.

Journeys in multispecies mobile video ethnography with wearable cameras

The Cairngorms National Park (CNP) is the last remaining UK stronghold of the capercaillie – the world's largest grouse – but populations here have reached critically low levels at around 1000 birds. Recreation disturbance has now joined habitat loss, climate change and predator control as the focus of managerial attention to reverse the fortunes of this protected species. Dogwalking has been specifically highlighted as having profound implications for the breeding practices and outcomes of capercaillie (Moss et al., 2014) and has attracted additional management interventions. Under Scottish outdoor access law, recreation can only be undertaken "responsibly". This means keeping dogs under "proper control" and being "on a short lead or close at heel during the breeding season (usually April to July) in areas where there are ground nesting birds breeding and rearing their young" (SNH, 2015, p. 15). Our case study focused on a peri-urban woodland in the CNP where the first significant targeted management initiative attempting to manage recreation for capercaillie objectives in the United Kingdom took place. Since 2012, the land managers specifically requested that dogs must be put on leads in the most "sensitive" area, marked by signposts and maps.

Sundberg (2006) identifies conservation projects as important contact zones that can be sites not only of "contestation and conflict, but also connection, empathy and contract" (239). Likewise, we found dogwalking amongst protected species to be an everyday, often taken-for-granted practice that involved multiple, interwoven more-than-human contact zones in which the thriving of capercaillie, dogs and humans were *all* at stake and negotiated in an ongoing, fleshy way through the making and disrupting of interspecies connections. Between 2012 and 2016, we enrolled wearable cameras into research encounters with local dogwalkers who regularly used the case-study woodland to explore how various ways of performing human-animal contact zones shaped the ability of capercaillie, dogs and humans to co-exist together. We especially wished to understand the aspects of embodied, mobile practice that mattered in whether cross-species connections or ruptures were generated, and in turn understand better the "crafting of shared lives" (van Dooren et al., 2016, p. 10).

We had already been experimenting ethnographically with WCT since the mid-2000s to explore encounters between different outdoor recreation users, principally walkers and mountain bikers, as a way to address limitations we experienced with "go-along" or "walk and talk" methods (Kusenbach, 2003). Although very valuable in immersing researcher bodies in encounters (after Longhurst et al., 2008), we found go-along methods tended to privilege verbal communication and discursive agency. They can even preclude the unfolding of the very practices the research sought to understand, especially with fast, technical or tricky practices in confined spaces. Enrolling a wearable camera on the human participant's body allows the researcher to have a screen-based multispecies encounter that is separated in time and space from the direct, fleshy encounter taking place between species in their habitats (see Figure 9.1). The camera can also be enrolled with bodies of nonhuman animals (see Figure 9.2). By generating audiovisual material through the movements of the camera-body, and subsequently audiencing that recorded material space is retained (both bodily and attentionally) in the original fleshy encounter for the conjoint doing of more-than-human bodies with their ecologies. Particular space is made for body language and other forms of nonhuman agency, such as the textures of terrain and the atmospheres of landscapes.

Figure 9.1 Image produced when camera is enrolled with human participant.

Figure 9.2 Image produced when camera is enrolled with a canine participant.

Our ethnographies with dogwalking[5] participants therefore evolved into the following three-stage process. First, we conducted a pre-outing, semi-structured biographical interview, usually in participants' homes, to build rapport and to situate how the dog-human relationship and dogwalking practices figured in the lives of the main human participant and of other household members.

Second, we organised a wearable-video-recorded outing where the human participant wears a head- or chest-mounted camera and goes for one of their typical dogwalks in the pinewoods with the capercaillie-related management restrictions. We had to decide about whether and how to involve the researcher in these outings and piloted both accompanied and unaccompanied WCT-recorded dogwalks. We saw advantages to the researcher accompanying the dogwalking partnership for the first portion of the route, partly to make sure the equipment was working appropriately,[6] and partly to allow a go-along sensibility of moving-with and feeling-with those particular beings and that particular terrain, so as to provide a basis for "making sense" of experience together later whilst watching the video. However, we were acutely aware that the researcher *not* being there allowed the human and dog to focus on each other, and wider ecologies, in ways much more akin to their everyday dog-walking entanglements. We therefore settled on a roughly half-and-half approach, with the researcher accompanying the participant to begin with and then peeling off to allow the participant to continue on their own way, and return to the trailhead through the forest via a different route.

Third, the researcher and human participant viewed the video together in a post-outing video review interview, reflecting, commenting and discussing as they watched, creating a further audio recording. Both humans[7] had a chance to watch the footage beforehand so they were ready to highlight aspects or moments of particular significance to them. Each took a turn to be in control of navigating through the unedited[8] footage, though typically the dogwalk took no more than an hour so we just watched it through and paused when there was a major point to make.

It is possible to understand these interconnected research spaces as attempts to engage three notable kinds of multispecies contact zone[9]:

1   The *contact zone in which different species meet in mutual somatic proximity*: bodies converging closely enough for at least one to register and respond (or not) to the presence and articulations of at least one of the others, for us the proximal, material, multi-sensory, affective meeting of canine, human and (potentially) avian bodies in the woods. This is the kind of contact zone of proximal intercorporeal affective relations that Haraway (2008) elaborated in relation to the agility training done with her own dog;

2   The *contact zone in which the camera was enrolled with and embodied by a particular mobile subject as it met other species in somatic proximity*: the intimate fleshy entrainment of imaging and sound-recording technology with a motile body, or potentially multiple bodies, makes it possible to generate audio and video traces of the encounter as it was performed and experienced by that particular body;

3   The *contact zone in which a participant views (and hears) the footage produced by the camera and has the opportunity to respond to it*: a foundational aim of orchestrating this screen encounter between a researcher, a participant (in this case both human) with audiovisual traces of the fleshy encounter that originally took place (in our case in the woods) is that the traces will act as something akin to a proxy contact zone for the fleshy encounter, or at least serving to evoke memories, thoughts,

sensations and feelings relating to the corporeal and affective of those more-than-human bodies as they entangled in another space and time. The human participant and researcher are thus afforded the opportunity to use the more-than-human traces to work towards a shared understanding and perhaps even mutually empathetic understandings of the dogwalking practices in question; attuning together, feeling together, knowing together. In our case the camera was worn – and the video was audienced – by human participants but hypothetically there is no inherent reason the generation and audiencing of audiovisual materially cannot more centrally include nonhuman or multiple creatures in the future, *if* frontiers of more-than-human understanding demand it, and *if* the implicated ethical relations and technological barriers are addressed. This is explored further in the section "*Who wears the camera (and how) matters*".

Making audiovisual traces of more-than-human animals with wearable cameras

Speaking "with" and "to" a body-camera in multispecies ethnography

We found that the expressive movement and stillness of more-than-human animals was translated into the audiovisual traces of a wearable video camera in two key, mutually constitutive ways. Not only were traces made by animals being present within, moving within, or moving in and out of "the frame" or more accurately the camera's field of view (FOV); in this case largely dogs but also other humans. Audio and visual traces were also made by the mobile practices of the animal wearing the camera unit (in this case humans). As highlighted by Bégin (2016), a distinctive trait of a wearable camera is its ability to record an articulation of the movement of the body or body part upon which it is mounted. The resulting agency of the wearing subject–object thus invites a kind of incidental filming, which is distinct from typical handheld intentional shooting in that the lens is not moved in specific efforts to frame and capture the doings of particular species or ecologies but is moved in response to the ways in which the wearer co-performs those ecologies as an unashamedly involved participant. In our case the dogwalker-human was moving largely in response to their dog, the terrain, landscape and other forest users. The camera makes audiovisual traces of the pinewoods experienced as surfaces felt and used, sounds heard, and scenes seen. Since WCT "survives" and keeps up with movement that could break or elude other forms of research instrument (including researchers' bodies at times!), it could keep up with the human – and to a lesser extent dog – participants and make traces all the way from their home (or car) through the forest. The footage their head-cam or chest-cam shot was thus an expression of their lived practices as a continual flow of (generally mundane) encounters with particular species, materialities and ecologies.

It was striking how different the canine and avian (and indeed human) artic-ulations were in terms of what was made available to the camera in terms of their spatial, social, sensorial and ecological priorities, as expressed through presences/absences, movement/stillness and the generation of associated affec-tive intensities and animal "atmospheres" (Lorimer et al., 2019).[10] The caper-caillie is a ground-nesting bird with offspring vulnerable to numerous predators – not just dogs being walked but also wild species such as foxes, pine marten and crows – and therefore concealment is a core adaptive strategy. It was under-standable, therefore, that this bird was completely absent in the camera's visual registers likely because it was always doing its best to hide in the undergrowth or in a tree (see Clip 7).[11] The only visual traces left by capercaillie were feath-ers, scat or dust bath areas which the small number of humans with the knowl-edge to recognise them stopped and stooped to examine them further, thus allowing the camera to "see" them too, sometimes purposely highlighting it for the researcher[12] (see Clip 8).[13] Our fieldnotes mention the very occasional aural traces of the sound of large wings flapping deeper in the forest, which could have been the sound of a capercaillie being "flushed" but the camera microphones were not able to pick up well sounds so far from the wearing body.[14] The bird's most significant visual trace was through the olfactory reg-isters expressed through canine bodies (of which more later). Yet through such conspicuous absence and indeed absent presences the bird articulated its sense of bodily jeopardy, fear and the need to hide any fleshy presences. If anything was being "said" by the capercaillie, it was their desire for a non-contact zone that was articulated through its intentional absences. In a sense, the vegetation was articulating too in terms of the work done to provide concealment.

In contrast, dogs articulated richly in ways the camera could register vis-ually. They were typically present, and stayed present, within the frame as they tended to trot a little way ahead of the human for most of the time they were off-lead, only occasionally dropping behind them to continue sniffing a particularly interesting spot, go to the toilet or running away to chase wildlife or interact with another dog. Furthermore, a large variety of canine move-ments were available to the camera's visual field, from the micro-gestures of a twitching nose or cocking an ear back to hear something behind, to the shape, direction, speed and rhythm of dog's wider ranging movements as they sought out, and followed, scent trails. Such animal mobilities are not abstract shifting spatial co-ordinates, however, but an inextricable embodiment of the lived experience of animals themselves (Hodgetts & Lorimer, 2018). Accordingly, the camera was able to make visual traces of movements that expressed something of the dog's interests and attachments, including efforts to affect, and invitations to be affected by, their human partner, other dogs or humans, wildlife, vegetation, terrain, and other materialities and atmospheres of the forest.

Three particular affective dynamics expressed through the movements enmeshing in three key modes of "dogness" were picked up by the camera whilst out on these walks: scanning, noticing and pursuit. The first was a kind of *monitoring, sweeping* and *scanning* for visual, aural and especially olfactory information of interest articulated by other creatures, typically involving trotting and sniffing in a zig-zag motion, with the motion of sweeping across and beyond the path together with smelling the air and vegetation inviting scent molecules of concern to find their way into the dog's nose to be detected (see Clip 1).[15] Implicated too were the kinaesthetic and proprioceptive registers of the dog to navigate and retain balance amongst brash, roots, and tangled vegetation that made it difficult for humans to follow off the path (see Clip 2).[16] The second dynamic available to the camera was the movements of the dog associated with *noticing* something important, such as catching a scent or hearing a rustle in the undergrowth, and *garnering more information* about it, such as the stiff posture, staring and long, deep sniffs of a dog trying to gauge what a creature might be and where it is located. The third was the decisive *responses* of dogs to their successful quests for information, which included the dog changing direction suddenly in intent close *pursuit* of more information (see Clip 3)[17] or suddenly accelerating away in *pursuit* of prey or to play with another dog, perhaps out of the visual field of the camera (see Clip 4[18]).

However, dogs' articulations did not seem to register much aurally with the camera. The audio track tended to be occupied rather wholly by the aural traces of the human wearing the camera (see Clip 5),[19] such as commands given to the dog (e.g. "lie down"), conversation with human companions, other forms of vocal expression (intentional and otherwise, e.g. a sign or gasp), the speed, depth and effort of breathing, and the sound of rhythmic footfalls or pedalling, as well as the rustle of clothing or the wind. Together with the moving images, the resulting footage provided traces of human motions articulating their relationship with terrain (their comportment, orientation and ways of finding and maintaining balance, including in response to slips, trips, obstacles or texture), and their practices of route-finding and way-making (include direction, pace, accelerations, decelerations, pauses, improvisations and what they were responding to). Indeed, some articulations of textures, gradients and other characteristics of terrain, weather and landscape were also registered in the WCT audiovisualities, particularly as they were registered conjointly by bodies and camera. Human motions associated with making bodily contact with dog or signalling to the dog sometimes featured in the audiovisual material (such as a click, whistle or hands pointing or moving to pat knees), as did those associated with particular emotional or somatic states (e.g. trembling, straining, fatigue), but the camera did not have available to it human facial expressions and many gestures. Otherwise, however, the camera detected many traces of human responses to dog articulations (e.g. verbal or tactile rewards) and any lack of them.

Who wears the camera (and how) matters

After Rose (2015), we need to be clear on why we are enrolling visualities as we are, including particular fields of vision. We found the same point holds in terms of who and what is registered aurally or "heard" by a camera (or not), and the various settings employed on the camera (see Concept Box 9.2). In our study there were distinct asymmetries between the species wearing the camera (humans) and those invited to be "in front" of the camera (dogs, birds and wider ecologies). In part this comes from a stark asymmetry of orientation towards lens and microphone. If enrolling only one wearable camera in a study, whoever wears it is most "heard" and least "seen". In our case, dogs (and various biota of trees and shrubs) featured as the main subjects in the visual frame whereas the humans dominated the audio recording. Some expressions of capercaillie could in theory have been registered by the camera, but the propensity of the bird to remain hidden from view meant on one hand that the camera "saw" capercaillie much like humans did (i.e. not really at all), and on the other hand that other technologies (e.g. thermal imaging) would arguably be needed to make the bird present in a way more akin to predatory species like dogs who could smell them even if they couldn't see them.

Concept Box 9.2:   Wearable camera settings (and how it might matter to the study of multispecies becoming)

Wearable cameras have a growing range of settings from which to choose. Even selecting the default "auto" settings is still a decision about choosing to make animals visible (or not) in particular ways and not others. Here are three important settings to illustrate why settings matter:

   Field of View (FOV): which on our GoPro, as with many action-cams, offers three options of "narrow" (90°), "medium" (130°, slightly fish-eye) and "wide", (170°, fish-eye). Each FOV option offers something different in terms of how wider ecologies have the chance to articulate, or how much attention is draw to what is in front, rather than to the side, of the camera and its wearer. We used the "wide" setting to give capercaillie the greatest chance of having their presences and absences registered, as well as the sinuously movements of most of the dogs.
   Frame rate: The greater the frames per second the faster the animal motion that can be registered without blurring. Very fast frame rates (e.g. over 100fps) allow the possibility of viewing the footage later in slow motion – a distinct way of seeing animals not available to the naked eye – but a setting that can only be used in favourable light conditions. We used 50 fps as this was enough to register the likely speed of the main species in question, and not so high that it would be a problem in the low light of the forest.

Colour: Different colour settings affect how close to a typical human way of experiencing colour the images are recorded (e.g. how green the trees are, how vivid subtle instances of colour are). This is another methodological decision that must be tailored to the specific research question. We used the "flat" setting as it is designed to be closer to the way humans register colour directly. The standard GoPro colour is a super-saturated grade that makes all colours much more vivid, which could affect the work the images do when viewed, and is thus an important methodological consideration.

Wearable cameras were developed purposely to evoke a (human-centric) point of view perspective and we must keep foremost in our minds that it is a specific and situated point of view evoked, including species-specific and individual-specific. It is important to think deeply about which species and individuals get to evoke their point of view with a camera and which are gazed upon, and how consenting they are to have their bodies recruited by one or both forms of visual engagement. Both positions have been identified as problematic and colonising regarding the exercise of power (Haraway, 1988, 2008). Moreover, we need to be aware of how the aural registers favour the *wearer's* articulations and the visual registers the non-wearers "in front" of the camera, and the implications for who or what is privileged and how. This includes how the cameras audiovisualities relate to the sensory hierarchies and somatic expression of the species in question. These in turn shape the nature and intensities of contact, mutual affect, and co-becoming between human and nonhuman animals that are made available to filmic registers.

Nevertheless, greater symmetry in recording video is possible. In principle, depending on the ethical relations and research questions involved, a researcher could have one or more nonhuman animals wear a camera.[20] Indeed, Seegert (2016) and commentators on *Leviathan* (e.g. Landesman, 2015; Stevenson & Kohn, 2015) underscore the value of WCT in de-centring human perspectives precisely *because* they can be worn by nonhumans or mounted at unusual angles. Is it, therefore, a problem when a study such as ours has human POV is at the heart of the camera enrolment, with all the possible privileging of human perspective that could entail in the ethnographic encounter? Is species asymmetry in camera-wearing necessarily a problem? The findings here suggest (a) that species symmetry in filming is already impossible because of asymmetries in species characteristics, most obviously ability to consent, whether the creatures are wild/tame, prey/predator or have the physiology to assimilate the weight and encumbrance of a camera without harm (see Bodey et al., 2018), and (b) that cameras worn by humans can still provide opportunities to register nonhuman and indeed more-than-human articulations and responses that might otherwise have been difficult.

Researchers just have to develop awareness of the forms of animal expression invited with the lens and microphone of a particularly embodied camera, and those we might miss, and the associated sensory and corporeal politics involved. For example, two forms of sensorial expression WCT has been specifically lauded for evoking – breathing and kinaesthesia (Spinney, 2011; Brown & Lackova, 2020) – are typically only relating to wearer of the camera. In our case this meant the nonhuman animals could not generate audiovisual traces of the forces, rhythms and patterns of their bodies' visceral movements and physical dialoguing with vegetation and textured terrain as intimately or immersively as if the camera had been more proximate with their bodies. The footage registered, therefore, aspects of the human's kinaesthetic and respiratory experience far more than the animals elsewhere in the filming assemblage. Comparing this with my very limited dog-cam experiments with my own dog (see Clip 6)[21] suggests this would be a loss. The olfactory priorities of dogs become clear in our main method but provide no chance to pick up important subtleties and affective qualities of these movements (e.g. I was struck by the sheer effort and relentlessness of all that sniffing). We also lose a visual sense of how human handlers make themselves available to dogs, for example through aspects such as the precise quality of eye contact. Ethnographic observation on the dogwalks I accompanied suggested that humans articulated and responded with their dog partners in a range of extralinguistic ways – intentionally and unintentionally – using "nonsymbolic representational modalities" (Kohn, 2013, p. 8). But many of these were not registered by the camera. Furthermore, our camera did not register any of the experience of capercaillie, such as how a female caring for her brood might sense and be affected by humans and predators coming closer through the vibration of feet/paws or sound of bodies brushing through vegetation. Questions are prompted of whether the fleshy vulnerabilities of different species are being evoked with adequate parity.

The point remains, however, that human POV filming made available novel audiovisual traces of canine and avian co-constitution of space, with the potential to invite novel ways of noticing, attending and perhaps even attuning and caring. There is no inherent need for symmetry for symmetry's sake. As Van Dooren et al. (2016) remind us, "multiplying perspectives is not simply about assembling diversity, nor is it about the adoption of an easy relativism; rather, it is about 'staying with the trouble' in an effort to meaningfully navigate one's way through the complexity of worlds in process" (pp. 11–12). All we need is to be clear about our research question – regarding the creaturely co-becomings we most want to know about and why – and what we hope to get from playing with familiarity and strangeness when we evoke being "in the shoes" of another, whether that is one of our own species or not.

Conclusion: mobilising multispecies video encounters as key contact zones

Asking the right questions of, and with, animals (Despret, 2016) in order to take seriously questions of agency, subjectivity and "voice" is at once an endeavour of asking the right way. Doing MVE with wearable camera technology (WCT) is one of an increasing range of ways being experimented with to heed the call for better critical animal methodologies (Buller, 2015) and invite animals to "speak" more on their own terms (Birke, 2014). In our experience of multispecies ethnography, enrolling a wearable video camera opens up many fresh possibilities for registering how animals "speak", especially how human and nonhuman animals articulate to each other through embodied, mobile practice in multiple, interwoven contact zones; these places of "spatial and temporal copresence" (Pratt, 1992, p. 7) where "most of the transformative things in life happen" (Haraway, 2008, p. 219). This chapter has elaborated on the kinds of multispecies articulation and response that were made available to our wearable video camera in the human-dog-capercaillie contact zones of our study.

In common with film techniques more generally, WCT can generate audiovisual traces that are highly inclusive of the more-than-verbal, multi-sensory, somatic forms of expression so significant to nonhuman animals. The distinct additional strength of WCT is enabling researchers to take seriously the methodological implications of multispecies bodily expression continually coming into being with specific mobilities and ecologies. Being better able to witness and evoke "animals' experiences of (im)mobility" (Hodgetts & Lorimer, 2018, p. 9) – and the related, power-laden opportunities and frictions of distance, speed, effort, rhythm, direction and predictability in creatures' intersecting trajectories – is increasingly recognised as a hitherto neglected area of human-animal studies. I would suggest too that it is especially underexamined regarding mundane, taken-for-granted multispecies mobile practice.

The capacities of WCT to attach to bodies, and remain enmeshed with them whilst moving, enables angles, proximities and immersions in the thick of the action of contact zones, which open up possibilities for engaging with a multitude of unfolding and enfolded spaces and times of human-animal research encounters that would otherwise remain elusive, or perhaps be too intrusive, intimate or quotidian. Uniquely, they register the incidental movement of the camera wearer as well movements of other species available to the field of view. Such a mobile, embodied approach is particularly sensitive to the flux, flow and spatiotemporal patterning of animal lives; their range, journeys, comportments and gestures, their pulse, pace and periodicity, their kinaesthetic and olfactory expressiveness and their fleshy vulnerabilities. This means when we use WCT we can work harder to stay with animals in their dynamic, multi-sensory and nonverbal world, and engage audiovisualities in ways that allow them to articulate more on their own terms and express their own practices of meaning-making (after Kohn, 2013).

Crucially, if multispecies becomings are inextricable from their constitution through myriad mobile practices (Hodgetts & Lorimer, 2018), then highly entangle-able audiovisual technologies like wearable cameras offer potential to stay with more of the "trouble", for more of the time, or at least generate more traces of the more-than-human co-becomings through which trouble is made, averted manifest and experienced in fleshy, creaturely lives.

Previously ethnographic attempts to register animal becomings with static tripod-mounted or hand-held cameras, or with the limitations of the human body, meant researchers only stayed with the trouble that was most conveniently stayed with, arguably missing or marginalising a whole swathe of dynamic and somatic multispecies practices and experiences that could be vitally relevant. But *which* troubles and *whose* troubles are we endeavouring to stay with? Sundberg (2006) underlines how the specificity, situatedness and embodiment of the human-animal encounter matter for who gets to make and avoid contact in particular contact zones, how, and which knowledge practices, species and individuals are privileged.

Enacting such novel ways of seeing depend on precisely how we enrol bodies, cameras and environments in both the making and the audiencing of the images and sounds (Brown & Lackova, 2020). A contact zone already often involves "highly asymmetrical relations of power" (Pratt, 1991, p. 33) and these can be further exacerbated or ameliorated depending on how camera technologies and techniques are enrolled. Taking on board the insights of Haraway (1991) and Rose (2015) means paying attention to which human and nonhuman animals are made visible, invisible, and how when we enrol particular types of video. Our research suggests the precise embodiment of a wearable camera – in terms of who wears it and where it is positioned – matters fundamentally for the animal articulations available to the lens and microphone, and thus how particular species and individuals are "seen" and indeed "heard" (and which are absent) through that subsequent audiencing of filmic media.

In our study there was profound asymmetry amongst the creatures invited to co-generate audiovisual materials in the MVE. Only humans wore cameras and thus humans, dogs, birds and wider ecologies had differing roles and agencies in authorship, and were differently visible – and, crucially, audible – in the footage recorded. There is no doubt too that an element of speciesism exists due to video images favouring more animate species or species that articulate most like humans. Can we argue that the enrolment of WCT in particular contact zones help avoid more harmful contact in other zones? And how do we deal with the huge range of creatures co-mingling in these contact zones beyond a narrow focus on the proximal encounters of a handful of species?

Here it is useful to recall that in Pratt's and Haraway's notion of contact zone, asymmetrical power does not mean one party necessarily rolling over subordinately for the other. Likewise, the asymmetrical evocation of a mainly human "point of view" as a way in to particular multispecies contact zones still suggests elements that can be constructively transformative and generative. Animals can still exert agency despite their difference, and perhaps in more expansive ways than with established, typical discursive and video methods. By allowing animals to "speak" in broader terms that acknowledge the extralinguistic work bodies and movement do in multispecies becoming, video from a human perspective can still invite less-appreciated nonhuman and more-than-human articulations to be registered can assist in working towards re-centring nonhuman animal agency, experience and subjectivity. Furthermore, there are whole worlds of experimentation to be explored through the manipulation of video footage – such as playing around with temporality (e.g. frame rates and playback speeds), colour, field of view – and/or interweaving video with other ways of visualising or making tangible creaturely articulations, including the realms of the biochemical, electromagnetic, supersonic, cellular, and so on (see Hodgetts & Lorimer, 2015). These are surely worth exploring further as long as we retain our finger on the pulse of the politics of visuality.

Whilst exploring the terrain of what wearable cameras make possible and available to the multispecies encounter, we must remain acutely attentive to the ways in which visualising animals can have negative implications for them. Only then can we make a call on whether a particular embodiment of a camera within a species assemblage is an exploitative colonisation, as asserted by Haraway (2008) or an opportunity to better express nonhuman agency in cross-species knowing, as suggested here. Therefore, extending a point made by Duru (2018), I suggest that we keep the mobilisation of video in radical more-than-human ways as an open question deserving of further exploration.

Ultimately, enabling animals to "speak" is not powerful unless they are also "heard". Visual technologies such as WCT may invite and allow nonhumans to articulate more on their own terms, but this does not guarantee humans are noticing, listening or attending any better. We need to pay more attention to the audiencing of multispecies film and video in ethnographic practice, for example, how particular audiovisualities shape affective relations and practices of noticing and attending.

Notes

1 Traditionally ethology has typically sought to generate objective accounts of animal lives as if self-evident reality abstracted from human influence, with little critical reflection on how the practices of animals are co-constituted with humans and human enrolments of moving images.
2 We humans share with other creatures 'indexical' semiotic modalities which involve "signs that are in some way affected by or otherwise correlated with those things they represent" (Kohn, 2013, p. 8).
3 These authors discuss 'Leviathan' (2012), an experimental documentary film by Castaing-Taylor and Paravel shot entirely with GoPro cameras, which provides an immersive portrayal of the more-than-human relations of deep-sea fishing practices.
4 See Brown et al., 2008; Brown & Spinney, 2010; Brown & Dilley, 2012; Brown & Banks, 2015; Brown, 2015, 2017).
5 Of the 25 dogwalkers, 10 were doing MVE and 15 doing Participatory Video, and were in addition to 24 other recreationists and 14 non-recreationists covering the main stakeholder types.
6 Until the advent of the GoPro 'action-cams' were notoriously unreliable (e.g. moving position, cables falling out and or difficult to set up due to lack of viewable viewfinder to check the images the camera was recording).
7 The researcher watched the uncut footage through in real-time between Stage II and III, taking fieldnotes, attending, for example, to the happenings (and

non-happenings) being witnessed and any feelings, sensations, thoughts, impressions evoked by the viewing. Care was taken to think critically about aspects of the researcher's background, the framing of the research questions and protocol, and of broader visual cultures and their affective logics, and how they might be shaping the feelings and understandings being generated.

8 This was because we did not want what the researcher counted as a 'happening' or notable – as inevitably would be foregrounded in the editing – to obscure what was important for (and emerging as interesting about) the participant. This included the mundane and taken-for-granted aspects of dogwalking practice that neither participant nor researcher would necessarily have foreseen as important until they made the space to sit together contemplating it, witnessing it and enabling it to affect them.

9 These are contact zones conceived not as discrete spaces but key junctures in multispecies networks that provide a site of – and way into – embodied ethnographic engagement.

10 Inferring the mobilities and experiences of animals from observation "requires care and reflexivity. Movement can be misleading, and misinterpretation is always a possibility in multi-species ethnographic research (Kirksey & Helmreich, 2010)" (Hodgetts & Lorimer, 2018, p. 10). Cautions that apply to direct observation of animals in situ are even more strongly applicable when viewing video of the same.

11 Katrina Brown (2019, September 12) *Ground-nesting birds are hidden in the field of view* [Video]. YouTube. https://youtu.be/8BOCC8i2RAc

12 This only happened with those participants who also considered themselves wildlife enthusiasts and occurred either when researcher was present in the go-along half of the outing or when the participant 'spoke' to the researcher through the camera.

13 Katrina Brown (2019, September 12) *Spotting capercaillie poo* [Video]. YouTube. https://youtu.be/wkf_MLYgMPk

14 In the context of bird life, the term 'flush' means to disturb a bird in such a way that they are driven from the position they were occupying. In the case of capercaillie this would likely involve movement between nest, ground and trees.

15 Katrina Brown (2019, September 12) *Sweeping, scanning, sniffing* [Video]. YouTube. https://youtu.be/7xR61fbdb8E

16 Katrina Brown (2019, September 12) *Dog navigating brash* [Video]. YouTube. https://youtu.be/du51dtPv7Jk

17 Katrina Brown (2019, September 12) *Dog responsive to sensory experience* [Video]. YouTube. https://youtu.be/kLqR-em7z6Q

18 Katrina Brown (2019, September 12) *Dog leaving the frame* [Video]. YouTube. https://youtu.be/IwOZBOwpA64

19 Katrina Brown (2019, September 12) *Human-made sounds dominate the audio* [Video]. YouTube. https://youtu.be/W8ATjbO-jAk

20 There were practical and ethical reasons for sticking with human-cam on this occasion. On one hand, it was only in the latter stages of this study that a dog-cam harness became commercially available by which time we were nearly finished fieldwork. The lead researcher did experiment with her own dog to establish how it might work in a research encounter and this fed into the mutlispecies-made film used in another project: https://www.hutton.ac.uk/research/projects/TRANSGRASS

21 Katrina Brown (2019, September 12) *Dog-cam* [Video]. YouTube. https://youtu.be/3kdXFaN9hH0

References

Adams, I., & Mastracci, S. (2017). Visibility is a trap: The ethics of police body-worn cameras and control. *Administrative Theory & Praxis*, 39(4): 313–328.

Bastian, M. (2017). Towards a more-than-human participatory research. In: M. Bastian, O. Jones, N. Moore, E. Roe eds. *Participatory Research in More-than-human Worlds*. London: Routledge. pp. 19–37.

Bastian, M., Jones, O., Moore, N., & Roe, E. (2017). More-than-human participatory research: Contexts, challenges, possibilities. In: M. Bastian, O. Jones, N. Moore, E. Roe eds. *Participatory Research in More-than-human Worlds*. London: Routledge. pp. 1–16.

Bear, C., Wilkinson, K., & Holloway, L. (2017). Visualizing human-animal-technology relations. *Society & Animals*, 25(3): 225–256.

Bégin, R. (2016). GoPro: Augmented bodies, somatic images. In: D. Chateau, J. Moure eds. *Screens*. Amsterdam: Amsterdam University Press. pp. 107–115.

Birke, L. (2014). Listening to voices. On the pleasures and problems of studying human-animal relationships In: N. Taylor, R. Twine eds. *The rise of critical animal studies: From the margins to the centre*. London: Routledge. pp. 71–87.

Bodey, T. W., Cleasby, I. R., Bell, F., Parr, N., Schultz, A., Votier, S. C., & Bearhop, S. (2018). A phylogenetically controlled meta-analysis of biologging device effects on birds: Deleterious effects and a call for more standardized reporting of study data. *Methods in Ecology and Evolution*, 9(4), 946–955.

Brown, K., & Dilley, R. (2012). Ways of knowing for 'response-ability' in more-than-human encounters: The role of anticipatory knowledges in outdoor access with dogs. *Area*, 44(1), 37–45.

Brown, K., & Spinney, J. (2010). Catching a glimpse: The value of video in evoking, understanding and representing the practice of cycling. In: B. Fincham, M. McGuinness, L. Murray eds. *Mobile methodologies*. Aldershot: Ashgate. pp. 130–151.

Brown, K. M. (2015). The role of landscape in regulating (ir)responsible conduct: Moral geographies of the 'proper control' of dogs. *Landscape Research*, 40(1), 39–56.

Brown, K. M. (2017). The haptic pleasures of ground-feel: The role of textured terrain in motivating regular exercise. *Health & Place*, 46, 307–314.

Brown, K. M., & Banks, E. (2015). Close encounters: Using mobile video ethnography to understand human-animal relations. In C. Bates ed. *Video methods: Social science research in motion*. London: Routledge. pp. 95–120.

Brown, K.M., Dilley, R., & Marshall, K. (2008). Using a head-mounted video camera to understand social worlds and experiences. *Sociological Research Online*, 13(6), 1–10. http://www.socresonline.org.uk/13/6/1.html

Brown, K. M., & Lackova, P. (2020). Mobile video methods and wearable cameras. In: P. Vannini ed. *Handbook of ethnographic film and video*. London: Routledge.

Buller, H. (2015). Animal geographies II: Methods. *Progress in Human Geography*, 39(3), 374–384.

Collard, R. C. (2016). Electric elephants and the lively/lethal energies of wildlife documentary film. *Area*, 48(4), 472–479.

Despret, V. (2004). The body we care for: Figures of anthropo-zoo-genesis. *Body & Society*, 10(2–3), 111–134.

Despret, V. (2016). *What Would Animals Say If We Asked the Right Questions?* Minneapolis: University of Minnesota Press.

Dowling, R., Lloyd, K., & Suchet-Pearson, S. (2017). Qualitative methods II: 'More-than-human'methodologies and/in praxis. *Progress in Human Geography*, 41(6), 823–831.

Duru, A. (2018). Wearable cameras, in-visible breasts: intimate spatialities of feminist research with wearable camcorders in Istanbul. *Gender, Place & Culture*. 25(7), 939–954.

Fijn, N. (2007). Filming the significant other: Human and non-human. *Asia Pacific Journal of Anthropology*, 8, 297–307.

Fijn, N. (2012). A multispecies etho-ethnographic approach to filmmaking. *The Humanities Research Journal*, 18, 71–88.

Goode, D. (2006). *Playing with my dog Katie: An ethnomethodological study of dog-human interaction*. Lafayette IN: Purdue University Press.

Grandin, T. (2006). *Thinking in pictures: My life with autism*. New York: Vintage.

Greenhough, B., & Roe, E. (2011). Ethics, space, and somatic sensibilities: Comparing relationships between scientific researchers and their human and animal experimental subjects. *Environment and Planning D: Society and Space*, 29, 47–66.

Grimshaw, A. (2011). The bellwether ewe: Recent developments in ethnographic filmmaking and the aesthetics of anthropological inquiry. *Cultural Anthropology*, 26(2), 247–262. DOI: 10.1111/j.1548-1360.2011.01098.x

Hamilton, L., & Taylor, N. (2017). *Ethnography after humanism: Power, politics and method in multi-species research*. London: Palgrave Macmillan.

Haraway, D. (1988). Situated knowledges: The science question in feminism and the privilege of partial perspective. *Feminist studies*, 14(3), 575–599.

Haraway, D. J. (1991). *Simians, cyborgs, and women: The reinvention of nature*. London: Free Association Books.

Haraway, D. J. (2008). *When species meet*. Minneapolis: University of Minnesota Press.

Haraway, D. J. (2016). *Staying with the trouble: Making kin in the Chthulucene*. Durham, NC: Duke University Press.

Hodgetts, T., & Lorimer, J. (2015). Methodologies for animals' geographies: Cultures, communication and genomics. *Cultural Geographies*, 22(2), 285–295.

Hodgetts, T., & Lorimer, J. (2018). Animals' mobilities. *Progress in Human Geography*. DOI: 10.1177/0309132518817829

Kirksey, S. E., & Helmreich, S. (2010). The emergence of multispecies ethnography. *Cultural anthropology*, 25(4), 545–576.

Kohn, E. (2007). How dogs dream: Amazonian natures and the politics of transspecies engagement. *American Ethnologist*, 34, 3–24.

Kohn, E. (2013). *How forests think: Toward an anthropology beyond the human*. Oakland: University of California Press.

Konecki, K.T. (2008). Touching and gesture exchange as an element of emotional bond construction. Application of visual sociology in the research on interaction between humans and animals. *Forum Qualitative Sozialforschung/Forum: Qualitative Social Research*, 9(3), Art. 33. http://nbn-resolving.de/urn:nbn:de:0114-fqs0803337

Kusenbach, M. (2003). Street phenomenology: The go-along as ethnographic research tool. *Ethnography*, 4(3), 455–485.

Landesman, O. (2015). Here, there, and everywhere: Leviathan and the digital future of observational ethnography. *Visual Anthropology Review*, 31(1), 12–19.

Longhurst, R., Ho, E., & Johnston, L. (2008). Using 'the body' as an 'instrument of research': kimch'i and pavlova. *Area*, 40(2), 208–217.

Lorimer, J. (2010). Moving image methodologies for more-than-human geographies. *Cultural Geographies*, 17(2), 237–258.

Lorimer, J., Hodgetts, T., & Barua, M. (2019). Animals' atmospheres. *Progress in Human Geography*, 43(1), 26–45.

Moss, R., Leckie, F., Biggins, A., Poole, T., Baines, D., & Kortland, K. (2014). Impacts of human disturbance on capercaillie Tetrao urogallus distribution and demography in Scottish woodland. *Wildlife Biology*, 20(1), 1–19.

Ogden, L. A., Hall, B., & Tanita, K. (2013). Animals, plants, people, and things: A review of multispecies ethnography. *Environment and society*, 4(1), 5–24.

Paterson, M., & Glass, M. R. (2018). Seeing, feeling, and showing 'bodies-in-place': exploring reflexivity and the multisensory body through videography. *Social & Cultural Geography*. DOI: 10.1080/14649365.2018.1433866

Power, E. (2008). Furry families: Making a human–dog family through home. *Social & Cultural Geography*, 9(5), 535–555.

Pratt, M. L. (1991). Arts of the contact zone. *Profession*, 91, 33–40.

Pratt, M. L. (1992). *Imperial eyes: Travel writing and transculturation*. NY: Routledge.

Rose, G. (2015). *Visual methodologies: An introduction to researching with visual materials*. London: Sage publications.

Seegert, N. (2016). Dogme productions, with an emphasis on the dog: Revealing animal perspectives. *Film Criticism*, 40(2). https://quod.lib.umich.edu/f/fc/13761232.0040.209/--dogme-productions-with-an-emphasis-on-the-dog-revealing?rgn=-main;view=fulltext

SNH (2015). *Taking the lead: Managing access with dogs to reduce impacts on land management*. https://www.nature.scot/sites/default/files/2017-06/Publication%202015%20-%20Taking%20the%20Lead%20-%20Managing%20access%20with%20dogs%20to%20reduce%20impacts%20on%20land%20management.pdf [accessed 21.1.19]

Spinney, J. (2011). A chance to catch a breath: Using mobile video ethnography in cycling research. *Mobilities*, 6(2), 161–182.

Stein, R. L. (2017). GoPro occupation: Networked cameras, Israeli military rule, and the digital promise. *Current Anthropology*, 58(S15), S56–S64.

Stevenson, L., & Kohn, E. (2015). Leviathan: An ethnographic dream. *Visual Anthropology Review*, 31(1), 49–53.

Sumartojo, S., & Pink, S. (2017). Empathetic visuality: GoPros and the video trace. In: Cruz, E.G., Sumartojo, S., Pink, S. eds. *Refiguring techniques in digital visual research*. London: Palgrave Macmillan. pp. 39–50.

Sundberg, J. (2006). Conservation encounters: Transculturation in the 'contact zones' of empire. *Cultural Geographies*. 13(2): 239–265.

Van Dooren, T., Kirksey, E., & Münster, U. (2016). Multispecies studies: Cultivating arts of attentiveness. *Environmental Humanities*, 8(1), 1–23.

Vannini, P., & Stewart, L. M. (2017). The GoPro gaze. *Cultural Geographies*, 24(1), 149–155.

Wratten, S. D. (1994). *Video techniques in animal ecology and behaviour*. New York: Springer-Verlag.

10 Getting visceral

Body mapping the humanimalian

Heide K. Bruckner

Introduction

In *We are the Weather: Saving the Planet Begins at Breakfast*, author Jonathan Safran Foer struggles with the dilemma of how so many people accept the reality of a changing climate, yet at the same time feel completely unmoved, or emotionally uninspired, to take urgent action. Foer personally reflects on various aspects of this disconnect but centres on one key theme: how our habits, and the sensory encounters resulting from our habits, become something we unreflexively take for granted. In one vignette, he describes the feeling of coming home after a long trip and noticing the smell of your house, as if for the first time. Yet, after a few days, the scent dissipates – not through an air freshener or a change in the scent itself but through a sensory adaption of becoming accustomed to the scent. This sensory adaptation extends beyond smell:

> Sensory adaptation applies to all the senses. People who live beside construction sites tend not to hear the racket. When you rest your hand on a dog, you at first feel the warmth and fur, but after only a few moments you become unaware of touching anything at all. The sky is in my field of vision for most of the day, but other than when I deliberately focus my attention on something—a daytime moon, a rainbow—I am capable of forgetting that the sky is there at all. What is always there, stops being there… [it is what] we are least capable of accurately perceiving.
>
> (Foer, 2019, p. 112)

Through our everyday settings and environments, we have a myriad of experiences which blend into the backdrop of forming the rhythm, pace, march of our lives. Multiple sensory dimensions of the everyday get lost, at least in our ability to perceive them. What often also gets assimilated into the unreflexive backdrop of our world is the *presence of animals*: the squirrels on the telephone line, the fish in our water reservoirs, the moths flitting towards a porch light, the ground beef in our spaghetti sauce. Yet, these nonhuman actors are constantly part of our more-than-human world, and we learn, in their everydayness, NOT to see, hear, smell, or feel them. While Foer argues that a

DOI: 10.4324/9781351018623-13

re-sensitisation to our sensory worlds is a critical step towards climate change action, in this chapter, I argue that re-sensitisation to the animal presence in our world is a key feature for less exploitative relationships.

For human-animal scholars seeking to expand the humanities to consider the more-than-human components of our world, the challenge lies in how to take the everydayness of animals into account. How can we as researchers focus on "what is always there" – the sky, the scent in your apartment, the frozen chicken wings in the grocery store – while calling for a change in how these other-than-human components of the world are taken seriously, respected and cared for?

It is in response to these grand questions that I introduce a modest intervention in the methodological toolkit for human-animal scholarship: body mapping. In this chapter, I describe body mapping and argue that its application to (other) social sciences can be broadened to consider the animal influences on–and components of–our more-than-human bodies. I situate body mapping as a device to probe the "visceral" realm of the body, a realm I argue tackles both the personal and societal politics involved in our relationships with animals (that we eat). I draw specifically on body mapping research conducted in focus group settings around the topic of "humanely raised" meat, with detailed explanations on conducting body mapping and analysing data, and how the method could be appropriately applied for other human-animal research. Finally, I propose that body mapping can be utilised as a tool for human-animal scholars to get beyond the representational limits of scholarship in the social sciences. Instead, I assert body mapping is best understood as an intervention into the visceral realm, a material disruption to the sensory adaptation described by Foer above.

Through body mapping – and creating research opportunities that force a confrontation with the uncomfortable, difficult, or hard-to-perceive – scholars can emphasise the role that animals play as affective and material agents in our lives. Furthermore, I assert that body mapping as a method can serve as a way to intervene in our reality, not assuming that we as researchers merely perceive or capture existing human-animal relationships, but that through our research we *create* opportunities to explore the sensory realm of our relationships to animals.

The de-/re-sensitisation of meat as animal bodies

While Foer's sensory adaptation is certainly useful for thinking through how our quotidian habits accustom us to NOT hear, see, feel, or participate fully in a sensory engagement, it is worth iterating that for the case of eating animals, there are broader politics at play than just de-sensitisation (or sensory adaptation) on an individual level. For many in Western societies, meat has become so divorced from an association with the living animal that little concern is given to animal welfare in daily, mundane practices of meat-eating. As opposed to "forgetting" that meat comes from animals, several scholars argue that meat has been intentionally severed from animal origins in a myriad of discursive and material ways. From increasingly holding animals indoors away from sight (Joy, 2010) to the marginalised geographic locations of slaughterhouses themselves (Fitzgerald, 2010; Pachirat, 2011; Vialles, 1994), little political and societal importance is afforded to the over 70 billion land animals slaughtered each year for food (FAO Stat, 2020). Animals in meat become obscured through an elaborate process of linguistically re-naming, materially re-packing and re-assembling food products to bear scant semblance to their bloody origins (Buller & Cesar, 2007; Fiddes, 1992). When caricatures of animals do appear on packaging or advertisements–a serene cow grazing on an open meadow–these animal renderings appear as idyllic representations of animal husbandry rather than an accurate reflection of our contemporary intensive agricultural systems (Shukin, 2009).

That said, in recent years (especially in the EU, and to a lesser extent in the US), there has been a growing turn towards alternative food networks as a countermovement against the exploitative relationships of industrialised agriculture. These alternative food networks position themselves as "alternative" in several ways but often highlight a re-connection agenda of linking consumers to the origins of their food (for a summary of this research, see Goodman et al., 2012). Popular literature on ethical meat consumption highlights "knowing where your food comes from" (Pollan, 2006). However, critical food scholars critique this moral-cognitive model of ethical consumption, pointing out that the ways we experience food extends far beyond rational knowing. Instead, they emphasise that eating happens through multiple rational, sensuous and practical realms (Barnett et al., 2011; Eden et al., 2008). In the case of meat, alternative modes of meat production often foreground animal welfare and promote their meat originating from "happy" or "humanely-raised" animals. The notion of "happy" animals, codified both through marketing and through certification standards, exemplifies the contradictions inherent in sentient commodities (Wilkie, 2005). There is an obvious disjuncture between thinking (and feeling) animals as sentient, affective beings, while also positioning animal flesh as a commodity arising from an inanimate being.

Despite these obvious contradictions, human-animal scholarship has tended to eschew research on affective dimensions of meat eating within the "happy" meat framework. Human-animal studies that address farm animals as meat have done so in various ways, from the philosophical questions of eating meat (Adams, 1990; Potts, 2016) to biopolitical analyses of domesticating animals (Holloway & Morris, 2014) and worker-animal relations within industrial agriculture (Porcher, 2017). More specifically within literature on alternative food networks, human-animal scholars' engagement with concepts of animal welfare or "happy" meat has occurred within two primary frameworks: 1) critiquing animal welfare as a project of modernisation (Buller & Morris, 2003) or 2) regarding "happy" meat as a neoliberal greenwashing project (Stanescu, 2010). The first scholarly engagement focuses on standards and knowledge-making practices within animal welfare science and legislation, provoking compelling work on the affective dimensions of naturecultures in the socio-technics of knowledge (Latimer & Miele, 2013).

The other set of literature frames "happy" meat critically, citing minimally improved welfare standards and anthropomorphised portrayals of animals to argue that "happy" meat is inherently hegemonic and neoliberal (Cole, 2011; Stanescu, 2010). While not fully disputing that "happiness" is a quality of meat commodified and sold to consumers, I do argue that for those who purchase and consume "happy" meat, more nuanced human-animal relations are at play. Furthermore, by sceptically referring to meat as "happy" or "unhappy", the present literature avoids an in-depth discussion of the mechanisms through which shared well-being, or happiness, between farm animals and humans becomes palatable (however see Bruckner, 2018; Bruckner et al., 2019; and Miele, 2011 for notable exceptions).

By better understanding the motivations and experiences behind a growing movement of animal welfare certified and "happy" meat, we can glean clues on how conscientious consumers relate (and become re-sensitised) to the animals they eat. Donna Haraway refers to human-animal "'relatings" as a multi-layered term capturing a complexity of interactions and connections we have with animals. She forwards the concept of relatings to also capture a diversity of human-animal interactions, as well as inherent responsibility tied to sharing our worlds with animals. Specifically, in regard to eating meat she writes:

> There is no way to eat and not to kill. No way to eat and not to become with other mortal beings to whom we are accountable, no way to pretend innocence or transcendence or a final peace.
>
> (Haraway, 2008, p. 295)

Framing "happy" meat as simply a greenwashing technique, or a counter-movement to agricultural modernisation, is pretending transcendence or innocence – and absolving ourselves of the responsibility of acknowledging the intertwined human-animal worlds. Haraway advocates "staying with the trouble" and exploring the discomfort, responsibility and complexity of how, and why, people eat animals (Haraway, 2008).

Methodologically, how can one probe this accountability to (dead and live) animal bodies, and the sense of discomfort that may arise when asking meat-eaters to consider the animal origins of their meal? What methods can be effective in eliciting not only data on meat but also other forms of affective, embodied and (un)comfortable human-animal relations? And more broadly, how can methods in human-animal research re-sensitise us to the quotidian presence of animals in our daily lives? Faced with these questions, I turn to body mapping as a methodological entryway. The body mapping research from which this chapter draws occurred as part of a 2014–2018 study on small-scale meat production in Styria, Austria.

Before detailing the specific body mapping findings, and their relevance for other human-animal scholarship, I devote attention to the method of body mapping itself. I argue for its inclusion in the human-animal methodological toolkit, situating why, and how, mapping human bodies indeed extends existing social science methods to consider the relationships we have to/with other nonhuman bodies.

What is body mapping? Why map *human* bodies in human-animal research?

Body mapping is a qualitative method that involves research participants drawing images of their own bodies to visually represent, and reflect upon, aspects of their relationships to others. Participants are first asked to draw an outline of their body, and then use images, text or symbols to visually represent how their physical bodies (and emotional/physical selves) are impacted by their everyday experiences of the prompt at hand. After a pre-specified time period of individual body mapping (e.g. 20 minutes), participants then take turns verbally presenting their body-mapped images to the group. In the process, they share stories relevant to their depictions, thereby prompting group discussion on common themes, or contradictions, arising from the body maps. Thus, body mapping and presentations transition from a body mapping prompt into a broader focus group discussion.

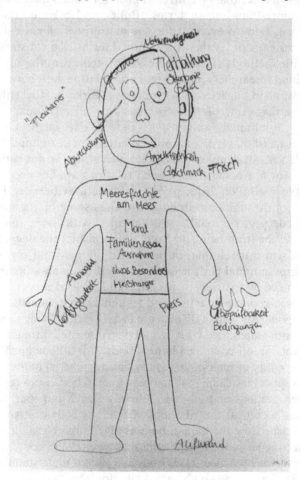

Figure 10.1 Photo by Heide K. Bruckner. Sample body map and quote: "You know that not everything in the industry is going well and that animals suffer at least in death, so that's why I have to shut it out—the emotions, the media, everything. Especially when my stomach rumbles".

While within research in the social sciences this method is fairly new, as a therapeutic tool, body mapping has been utilised for several decades (Cornwall, 1992; Solomon, 2002). For example, body mapping has frequently been employed as a tool to understand illness (Guillemin, 2004; Keith & Brophy, 2004; Solomon, 2002). It is now being used as a research method to understand the embodied experiences of undocumented migrant workers (Gastaldo et al., 2012, 2018), musical education (Griffin, 2014), youth sexual health (Lys et al., 2018) and HIV/AIDS activism in South Africa (MacGregor, 2009).

What connects body mapping research is the central importance of the body itself as a site of everyday emotional and material experience. However, there has been little scholarship on body mapping as a method to interrogate human-animal relations (Bruckner, 2018), especially in the current era where scholars (rightfully) push for understanding animals on their own more-than-human terms (Buller, 2015; Hamilton & Taylor, 2017). Indeed, how to account for the matterings of animals, either in relation to humans or on their own animal terms, is a constant source of debate. While methods selected necessarily stem from the topic at hand, human-animal researchers tend to focus on relationships between humans and animals, giving agency to both parties (Birke & Hockenhull, 2012). Buller states that "one of the most basic of all relational contexts for human and nonhuman animals, as indeed one might argue for all inter-species mixings, revolves around relations of consumption; of being made edible (or not being made edible), of eating (or not eating) and of being eaten (or not being eaten). Such relations are not necessarily fixed but assembled" (Buller, 2012, p. 51). Intersections between human and animal bodies through consumption are thus some of the more ubiquitous (and deeply personal) relationships with our animal world. Furthermore, far from being a "fixed" relationship, the degrees to which humans recognise the animal in meat, wilfully ignore it or eagerly consume a "happy" animal body arise from specific human-animal configurations, influenced by a variety of personal and societal expectations, bodily cravings, socio-economic factors and technocratic standards.

While respondents who body map are tasked with drawing their own bodies, what is both represented visually, and then discussed through storytelling, extends beyond a self-contained human body. It is a humanimalian body, a fluid, porous entity of intersecting symbolic and material human and animal worlds. To support my claim that mapping human bodies can get at the relatings between humans and animals, I briefly sketch out four primary and inter-related theoretical contributions of feminist body-centred scholarship which emphasise how the human body is always more-than-human. These four intersecting strands of scholarship include: 1) the body and ethics as shaped by everyday practice; 2) the body as a site for understanding biosocial relations; 3) the minded-body as a site for the "visceral" and 4) the ontological primacy of "becoming" advocated through body-centred research (see Jones' chapter, Chapter 4, in this book). After briefly reviewing each of these strands, I draw on examples from body mapping to illuminate how the method performatively (re)creates humanimalian bodies.

The body as a spatial site of research draws from feminist traditions of centring on *everyday practice* for data collection. Feminist political ecologists, in particular, have examined intersections of scale and power related to everyday engagement with animals, advocating for an ethics of care with/ towards animals that emerges from ongoing human-animal interactions (Donovan, 2006; Wolch et al., 2000).

While early scholarship in human-animal studies focused on animals as the Other, reflecting on animals as representations of what it means to be human (Philo, 1998; Wolch & Emel, 1995), there has been a shift to recognise the interdependence and relationality between human and animal bodies (Buller, 2015). The recognition that bodies are a site for *biosocial* relations establishes that the "outside" realm (or the Other animal body) becomes part of the "inside" embodied and emotional (human) realms through material and discursive intersections (Bull, 2009; Collard, 2012; Hovorka, 2012). Food scholars have long recognised the porous nature of categories of "outside" and "inside" the body, and draw attention to how eating animals brings relationality into view:

> Eating, as a conceptual and a physical act, brings foods and bodies into view; food does not remain on a supermarket shelf, in the kitchen or on a plate, but is placed in the mouth, chewed, tasted, swallowed and digested. Its solidity is thus broken down and rendered into fragments that both pass through, and become, the eater's body. This is a process that concomitantly establishes and ruptures social relations between bodies, whether those of the food's producers, retailers, micro-biological components or even of the original animal sources. Unpacking the encounters between foods (and human and nonhuman) bodies then offers a way to take account of the many networks and relations embedded in and performed by eating: these may be obscured by more established paradigms that consider food inside *or* outside the body, rather than inside *and outside* the body.
>
> (Abbots & Lavis, 2013, p. 1)

Through eating as everyday practice, scholars point out that bodies are *biosocial* relations. Such research understands "the body" as a lens through which to explore the social and biosocial relationships both within and *between* bodies – a decentred subject approach taking inspiration from material feminists (see e.g. Barad, 2012; Bondi, 2005; Grosz, 1995), actor-network social theorists (Latour, 2004; Law, 2002) and more-than-human feminist scholars

(Roe, 2006; Whatmore, 2002). A biosocial lens asserts that biological and social forces are internally combined, and thus feelings, experiences, or judgements are irreducible to being either/or biologically or socially motivated (Mansfield, 2008). However, while eating may be an obvious lens through which to apply the porous relationships of outside/inside the body, other material-affective biosocial relations in the way of multi-species becomings have been described by scholars who follow, for example, water voles (Hinchliffe, Kearnes & Degen, 2005), cows (Roe & Greenhough, 2014) or horses (Thompson, 2011).

Aside from every day and biosocial elements of the human body, which foreground relationality to other beings, contributions from visceral scholarship can provide fertile ground for human-animal studies. The burgeoning theoretical and methodological scholarship in visceral geography broadly engages with how bodies feel internally in relation to material and social space (Bruckner et al. 2021; Hayes-Conroy & Martin, 2010; Hayes-Conroy & Hayes-Conroy, 2010). This theoretical strand bridges divides between external/internal biological body and situates the body as inherently linked to cultural and economic structures. Furthermore, visceral scholarship can help "make a powerful link between the everyday judgements that bodies make (e.g. preferences, cravings) and the ethico-political decision-making that happens in thinking through the consequences" of everyday acts, like consumption (Hayes-Conroy & Hayes-Conroy, 2008, p. 462). Conceptualising the visceral as arising from a minded-body eschews the binary framework of mind v. body, external vs. internal; the visceral instead moves us towards understanding how eating bodies encounter food and moral decisions in everyday life. Harvey Neo, in a chapter about what motivates vegans to avoid animal products, similarly finds that both the "head and the heart" are central to the decision for people to become vegan (Neo, 2015). His findings echo earlier critiques about limitations of the rational decision-making framework for consumption practices. Instead, a minded-body, or an intertwined visceral framework, is better suited to address our relationships to eating animals. For human-animal scholars interested in relations arising from livestock and farmers, visceral methods have also been employed to understand emotional and rational attachments formed between farmers and their sheep (Holloway, 2001). It is this vein of visceral, affective human-animal relatings that body mapping finds its inspiration.

Concept Box 10.1: Visceral

Have you ever had a gut feeling about something? A certain intuition that is hard to explain? You might feel uncomfortable walking down a dark alley, in an unfamiliar city. Rationally, you might know that the crime rate is low, but there is still something that makes you nervous. You are experiencing a "visceral" feeling of fear. When something is visceral, there might not be an obvious rational explanation, but the feeling in your body is still real.

"Viscerality", or "the visceral" realm, refers to how our bodies *feel in relation to their surrounding social and material spaces*. What is central to visceral scholarship is the relationship between individual bodies and their broader social, cultural and political environments. As opposed to understanding bodies as individual or isolated entities, a visceral approach helps illuminate the patterns and societal structures that impact and are felt by the body (Hayes-Conroy & Hayes-Conroy, 2010). The body is thus the site where the individual experiences, contests or reshapes the broader social and economic patterns in their everyday lives.

Let's return to our example. The same dark alley might not feel scary for someone who lives on that city block or a person who works in the neighbourhood after dark. The visceral reactions to walking that alley might shift depending on your gender identity, your class, your race or your physical ability. While the material space of the dark alley is the same, the visceral reaction of a person might be different depending on their relationships to that space.

Geographers have taken interest in the visceral to understand embodied reactions to surrounding spaces. They are applying theoretical and methodological insight from visceral geography to get at the broader patterns of feelings of belonging, disgust, fear, racism, human-animal (dis)connection, food aversion or bodily motivation to eat "healthy" food (Bruckner, 2018; Joshi et al., 2015; Hayes-Conroy & Hayes-Conroy, 2010; Longhurst et al., 2009; Sweet & Ortiz Escalante, 2015). What has been missing is robust scholarship that applies visceral approaches to human-animal relations. The application of unpacking our visceral reactions are endless – and they leave open the possibility that with new methods, we might *feel* differently towards our more-than-human companions.

Finally, scholarship on both the visceral realm and the body more generally shares an understanding of the body as a site for non-deterministic, exploratory relationships which emerge out of the research process itself. Importantly, body-centred research places ontological primacy on subjects becoming rather than being (Carolan, 2011). That is, bodies are not represented but instead brought into being through practice, including research practice (Bruckner, 2018). Body mapping research conducted by Sweet and Ortiz Escalante emphasise the utility of body mapping in order to open up new realities for participants (Sweet & Ortiz Escalante, 2015). In their case, women in Colombia, Mexico, and New York City use body mapping and storytelling to highlight the embodied intersection of fear, gender-based violence and space. Beyond personal reflection, body mapping contributes to a new *collective* spatial knowledge (Sweet & Ortiz Escalante, 2015). The process of body map storytelling, or individuals presenting their body maps, does not only bring to light specific knowledges, but actually creates new insight through group discussion and reflexivity. Their conclusions echo other feminist and material scholars who draw attention to knowledge production through the research process itself (Asdal & Marres, 2014; Haraway, 1988). Broader literature in performative methodology takes seriously the role of objects and material engagement in creating lively forums for participation, and advocates for performative methods in nature-society research (see Lezaun, 2007; Marres & Lezaun, 2011) and in human-animal studies, more specifically (Geiger & Hovorka, 2015; Hovorka, 2015).

I similarly argue that the utility of body mapping to understand specific human-animal relations lies both in the individual body mapping reflections AND in the collective sense-making that occurs during performative storytelling and focus group settings (Bruckner, 2018). This insight relies on the notion of a more-than-human body shaped through everyday practice, comprised of intertwined emotional and rational realms, and in constant interaction with the broader biosocial (material, human and nonhuman) relationships.

Doing and analysing the humanimalian body maps

The body mapping data I draw on comes from seven focus groups conducted in Graz, Austria with participants in alternative food networks. For the study, I recruited via flyer, asking for participants who self-identify as "conscientious" meat consumers. I informed all respondents that participatory, visual methods would be utilised to explore meat-eating practices, thus priming them for the research activities. The participants, aged between 18 and 65, were divided in groups by similar age, with roughly even numbers of men and women. Focus groups were kept small to promote trust, and ranged from between 3–5 participants (total n = 27). Scholars emphasise that group composition, group size, and focus group location are key factors which can have an important influence on the nature of focus group discussion and activities (Hopkins, 2007). Especially when conducting a body mapping activity about issues which may provoke discomfort, a smaller group size can not only promote individual and group reflexivity, but also lend itself towards more intimate conversations about sensitive topics (Longhurst, 1996; Bruckner, 2018). Aside from focus group size and group composition, researchers should consider the role of location – ideally, a safe, neutral location at a time of day during which most participants are available–acknowledging that all decisions about research time and location will inherently exclude some potential participants

(Hopkins, 2007). In my case, focus group sessions with university students were conducted on campus in the afternoon, after classes, whereas sessions for older adults were conducted in the evenings. All focus group and body mapping activities were held in a conference room at the University of Graz, Austria.

Each research session began with an explanation of the prompt and the body mapping activity, reassuring participants that the artistic representations were meant as tools for discussion. Some research participants were at first a bit hesitant or self-deprecating about their artistic abilities, but I provided reassurance there was no "right" or "wrong" answers, and that they were free to opt out/only participate in the ways that they desired. Then, I allotted 20–30 minutes of individual body mapping time for participants to respond to the prompt through words, images and symbols related to their depicted silhouette. While I supplied large sheets of paper, coloured markers, crayons and pens, body mapping materials could well expand to other mediums like magazines for collages or even clay and tactile materials for sculpting/building 3-D representations. After individually body mapping, participants were asked to present their map to the group, and then we turned to a collective focus group discussion. To preserve data, the completed body maps were photographed, and all discussions were audio-recorded and then transcribed, with final transcriptions then translated from German into English[1]. Research participants were compensated monetarily for their time.

In analysing body mapped data, there are two critical components to consider: the themes represented through the body maps themselves, as well as the conversations that arise between participants during body map presentations. Code frames, or categories that sort body mapping data, can be a useful way to organise the major categories of analysis (Saldana, 2015). Sweet and Ortiz Escalante (2015) for example, develop body mapping code frames based on visual analysis of the body maps (Rose, 1995), as well as qualitative content analysis of speech and discourse (Schreier, 2012). While I shared their analytic goals of capturing both the visual images and text themselves as content, I sorted content into two categories of a) visual and speech representations depicted on the body map (what practices and emotions were represented) and b) thematic tendencies of connecting to/distancing from animals that arose in group conversation. I also took notes on how, or if, I noticed hesitancies, discomforts, or contradictions and what topics those emotional responses corresponded to. While each body mapping activity will need to identify the analytic categories of interest, visual, discursive and group discussion material can be rich fodder for code frame analysis.

From this analysis, body mapping highlighted several aspects of embodied human-animal relationships, and contradictions, in the process of buying, preparing and eating meat. Body mapping brought out the everyday, visceral and biosocial, illuminating the utility of body mapping to probe the more-than-human aspects of bodies. For instance, several participants drew comments next to their feet, indicating that they try to source humanely-raised meat, but not if they have to walk too far. Thus, they linked the spatial and practical considerations with their body – and the effort required – in order to put into action their ethical orientations.

Within a typical focus group conversation, several researchers draw attention to difficulties as participants seem surprised or ill-equipped to discuss animals within the context of meat consumption (Joy, 2010; Miele & Evans, 2010, p. 183; Neo, 2015). However, after body mapping, my research showed that various participants seemed primed for discussion after having individual time to body map. Several respondents drew explicitly the handling of meat as a way to connect the tactile dimensions of cooking with a concern for animal life:

> For a while I was vegetarian and then not, and at first it was hard for me to cook meat. Since then, though, I find it really important to touch meat, that you look at it yourself and process it yourself. Try it out with a pig's head, for example. I find that generally for meat eating it's important that you know what you're eating, where it comes from, and that you don't just have a cooked piece of meat lying in front of you. You should really learn what it is, where it comes from on the animal, how do I cut it from the bone, and then really get your hands dirty. That's why I think the haptic dimension is so important, as a way to connect to the animal.
>
> (Body mapping participant)

By "getting their hands dirty", this participant expressed a desire "not to pretend innocence" (Haraway, 2008), but instead face responsibility for the animal itself. Due to the prompt of focusing on the embodied aspects of conscientious consumption, participants are re-directed to consider the ways they already touch, and are touched, by animal life.

Moreover, body mapping foregrounds the contradictory and intertwined rational and sensory experiences of "conscientiously" eating animals. Body mapping data pointed participants to the visceral realm of human-animal relations as a complex, minded-body, relationship. Many become aware of their minded-body knowledge through discussing their body map representations:

> I think that the head is really important in all of this, to know where the meat comes from, when you read or see a label, that all goes through your mind about what it means about animal life. …But when you eat meat, and you are judging its quality, and the eyes, nose, mouth all play a role. The same goes for taste and how it sits in your stomach. I drew a stomach which doesn't have to do too much with my knowledge. Well, actually it is its own type of knowledge, like if something sits well or doesn't sit well in your gut.
>
> (Body mapping participant)

While the participant perhaps did not initially conceive of the gut as a source of knowledge, in presenting the body map, he comes to a conclusion that the visceral indeed holds its own unique insight.

Finally, body mapping highlights the ontological primacy of an unbounded, constantly shifting humanimalian body. As opposed to conceiving of body map depictions as fixed or stabile "representations" of actually-existing bodies, I take the performativity of research methods seriously (Asdal & Marres, 2014), asserting that research methods intervene and bring to light certain realities, ultimately (re)making our humanimalian bodies. Through storytelling and group discussion of body maps, participants are guided through both individual and group reflexivity (Bruckner, 2018). In the dialogue in which they challenge one another, come to new insights, or realise similarities in their body maps, participants find new ways of sensitising themselves to the animal bodies they eat. In the exchange below, participants shift how they express the relationship between animal, knowledge and taste:

PARTICIPANT 2: Yeah, okay, like Participant 1 described, I also drew my head as knowing where it comes from. It's how I personally decide where to buy meat from, like is it coming from the region or not. I don't know if I'm just making it up, but I find that meat from the butcher tastes better to me, than just meat from a random place, or if I buy it from someone I know, then it just tastes better or something.

PARTICIPANT 3: No, I think it's true for me, too. It just tastes better if I know where it's coming from.

PARTICIPANT 4: Really? Do you think if we did a blind taste test here, you would know from the taste which meat is locally sourced?

PARTICIPANT 3: Well, maybe not. But if I know that the meat is coming from a local farm, then I think about its quality differently and, I don't know, I might value it differently or try to be more aware when I make that steak – like treat it as a special occasion.

What body mapping illuminates is the humanimalian body, its ethics and its taste, is at once a material, affective, and sensory engagement. Moreover, as participants share their body maps with one another, they reflect on similarities and differences, and even come to new understandings of what motivates their relationships to the animals they eat. If the performativity of methods is to be taken seriously, researchers must consider opportunities that arise through research settings, including body mapping, towards bridging the great divides between nature/culture or human/animal – categories that are too often considered disparate (Asdal & Marres, 2014).

Methodological implications for more-than-human worlds: beyond meat

Visceral scholarship holds great potential towards holistic approaches to the emotional and material relationships we have with our more-than-human world.

Body mapping serves as a method to foreground the embodied, everyday experiences we have with animals in our world. Instead of understanding body mapping as merely a way to recount relationships to animals, I have here argued that body mapping can help re-sensitise us to the comforts, and (dis)comforts, of these human-animal relatings. The diversity of humanimalian body maps, far from being mere representations, prompt new types of human-animal relations and understandings of our human-animal worlds through dialogue, disagreement and artistic and emotive renderings of these relatings through body map storytelling.

In this chapter, I have illustrated the utility of body mapping methods for human-animal scholarship in the context of conscientious meat consumption. However, body mapping, as an entryway into sometimes (un)comfortable relations, holds potential for other critical human-animal scholarship as well. Some other applications could expand on the conflicted nature of animal bodies as commodities, such as the relationships between fur and fashion, or the making of animals (and animal renderings) into pet food, medicine, or within laboratory research settings. Body mapping could also complement research on human-animal relatings which explicitly depend on shared physical contact, such as horses and trainers, hunting dogs and hunters or pigs as guides during the search for truffles.

Furthermore, in considering methods, and body mapping as an example to probe the visceral, it is important to note that as researchers, we can pay attention to the visceral even when NOT explicitly body mapping. Qualitative research always engages with emotions and affect on some level (even if unacknowledged). In a *Geoforum* debate on the practices of visceral methodologies, Allison Hayes-Conroy makes explicit this point:

> If we take seriously the politics of research (and fieldwork), we will recognize that our methods are always inescapably visceral. Whether conducting discourse analysis or cooking, dancing or otherwise sharing with collaborators/participants, embodied feelings shape the social and political processes inherent in our work.
>
> (2017, p. 51)

Making sense of the animal presence in our world is at once a rational, emotional, and, in the case of meat, a metabolic process. Too often categories of other nonhuman organisms as "killable", as raw material for fashion, or as serving other utilitarian human purposes, are easy fallbacks for scholarship in the social sciences. However, if we are to re-sensitise our research to animal presence in our world, and explicitly acknowledge animals as (willing or unwilling) agents in the research process, we need methods that account for the vitality and active presence of animals in the world. Probing this viscerality of human-animal relatings through body mapping is a novel intervention towards that end.

Acknowledgement

The research presented in this chapter was made possible through the "(Un) Knowing Food" project, 2014–2018, funded by the Regional Government of Styria, Austria.

Note

1 Translations from German to English are my own. Recognising the myriad of ethical and practical challenges inherent in translation (Temple & Young, 2004), I was careful to remain loyal to research participants' meanings and descriptions, while ensuring the text was clear to an English-language audience.

References

Abbots, E.J. & Lavis, A. (2013) *Why we eat, how we eat: contemporary encounters between foods and bodies*. London: Ashgate.

Adams, C.J. (1990) *The sexual politics of meat*. New York: Bloomsbury.

Asdal, K. & Marres, N. (2014) "Performing environmental change: the politics of social science methods," *Environment and Planning A*, 46(9): 2055–2064.

Barad, K. (2012) "On touching—the inhuman that therefore I am," *Differences*, 23(3): 206–223.

Barnett, C., Cloke, P., Clarke, N. & Malpass, A. (2011) *Globalizing responsibility: the political rationalities of ethical consumption*. London: Wiley-Blackwell.

Birke, L. & Hockenhull, J. (2012) *Crossing boundaries: investigating human-animal relationships*. Boston: Brill Publishers.

Bondi, L. (2005) "Making connections and thinking through emotions: between geography and psychotherapy," *Transactions of the Institute of British Geographers*, 30(4): 433–448.

Bruckner, H.K. (2018) "Beyond happy meat: body mapping (dis)connections to animals in alternative food networks," *Area*, 50(3): 322–330.

Bruckner, H.K., Colombino, A. & Ermann, U. (2019) "Naturecultures and the affective (dis)entanglements of happy meat," *Agriculture and Human Values*, 36(1): 35–47.

Bruckner, H.K., Westbrook, M., Loberg, L., Teig, E. & Schaefbauer, C. (2021). "'Free' food with a side of shame? Combating stigma in emergency food assistance programs in the quest of food justice," *Geoforum*, 123: 99–106.

Bull, J. (2009) "Watery masculinities: fly-fishing and the angling male in the south west of England," *Gender, Place and Culture*, 16(4): 445–465.

Buller, H. (2012) "Nourishing communities: animal vitalities and food quality" in Birke, L. & Hockenhull, J. (eds.), *Crossing boundaries: investigating human-animal relationships*. Boston: Brill Publishers, pp. 51–72.

Buller, H. (2015) "Animal geographies II: methods," *Progress in Human Geography*, 39(3): 374–384.

Buller, H. & Cesar, C. (2007) "Eating well, eating fare: farm animal welfare in France," *International Journal of Sociology of Food and Agriculture*, 15(3): 45–58.

Buller, H. & Morris, C. (2003) "Farm animal welfare: a new repertoire of nature-society relations or modernism re-embedded?", *Sociologia Ruralis*, 43(3): 216–237.

Carolan, M. (2011) *Embodied food politics*. Farnham: Ashgate.

Cole, M. (2011) "From "animal machines" to "happy meat"? Foucault's ideas of disciplinary and pastoral power applied to 'animal-centered' welfare discourse," *Animals*, 1(1): 83–101.

Collard, R. (2012) "Cougar figures, gender, and the performances of predation," *Gender, Place and Culture*, 19(4): 518–540.

Cornwall, A. (1992) "Body mapping in health RRA/PRA," *RRA Notes*, 16: 69–76.

Donovan, J. (2006) "Feminism and the treatment of animals: from care to dialogue," *Signs: Journal of Women in Culture and Society*, 31(2): 305–329.

Eden, S., Bear, C. & Walker, G. (2008) "Mucky carrots and other proxies: problematizing the knowledge-fix for sustainable and ethical consumption," *Geoforum*, 39(2): 1044–1057.

FAO STAT (2020) *Primary livestock*, viewed 14 April 2020, Available at: http://www.fao.org/faostat/en/#data/QL/metadata

Fiddes, N. (1992) *Meat, a natural symbol*. New York: Routledge.

Fitzgerald, A.J. (2010) "A social history of the slaughterhouse: from inception to contemporary implications," *A Human Ecology Review*, 17(1): 58–69.

Foer, J.S. (2019) *We are the weather: Saving the planet begins at breakfast*. New York: Farrar, Straus & Giroux.

Gastaldo, D., Magalhães, L., Carrasco, C. & Davy, C. (2012) "Body-map storytelling as research: methodological considerations for telling the stories of undocumented workers through body mapping", viewed 6 January 2020, Available at: http://www.migrationhealth.ca/undocumented-workersontario/body-mapping

Gastaldo, D., Rivas-Quarneti, N. & Magalhães, L. (2018) "Body-map storytelling as a health research methodology: blurred lines creating clear pictures," *Forum: Qualitative Social Research*, 19(2): 1–26.

Geiger, M. & Hovorka, A.J. (2015) "Animal performativity: exploring the lives of donkeys in Botswana," *Environment and Planning D: Society and Space*, 33(6): 1098–1117.

Goodman, D., DuPuis, E.M. & Goodman, M.K. (2012) *Alternative food networks: knowledge, place and politics*. New York: Routledge.

Griffin, S.M. (2014) "Meeting musical experience in the eye: resonant work by teacher candidates through body mapping," *Visions of Research in Music Education*, 24. Viewed 8 March 2020, Available at: http://www.rider.edu/;vrme

Grosz, E. (1995) *Space, time, and perversion: essays on the politics of bodies*. London: Routledge.

Guillemin, M. (2004) "Understanding illness: using drawings as a research method," *Qualitative Health Research*, 14(2): 272–289.

Hamilton, L. & Taylor, N. (2017) *Ethnography after humanism*. London: Palgrave Macmillan.

Haraway, D. (1988) "Situated knowledges: the science question in feminism and the privilege of partial perspective," *Feminist Studies*, 14(3): 575–599.

Haraway, D. (2008) *When species meet*. Minneapolis: University of Minnesota.

Hayes-Conroy, A. (2017) "*Critical* visceral methods and methodologies. Debate title: better than text? Critical reflections on the practices of visceral methodologies in human geography," *Geoforum*, 82: 51–52.

Hayes-Conroy, A. & Hayes-Conroy, J. (2008) "Taking back taste: feminism, food and visceral politics," *Gender, Place and Culture*, 15(5): 461–473.

Hayes-Conroy, A. & Martin, D.J. (2010) "Mobilising bodies: visceral identification in the Slow Food movement," *Transactions of the Institute of British Geographers*, 35(2): 269–281.

Hayes-Conroy, J. & Hayes-Conroy, A. (2010) "Visceral geographies: mattering, relating and defying," *Geography Compass*, 4(9): 1273–1283.

Hinchliffe, S., Kearnes, M. & Degen, M. (2005) "Urban wild things: a cosmopolitical experiment," *Environment and Planning D: Society and Space*, 23(5): 643–658.

Holloway, L. (2001) "Pets and proteins: placing domestic livestock on hobby-farms in England and Wales," *Journal of Rural Studies*, 17(3): 293–307.

Holloway, L. & Morris, C. (2014) "Viewing animal bodies: truths, practical aesthetics and ethical considerability in UK livestock breeding," *Social and Cultural Geography*, 15(1): 1–22.

Hopkins, P. (2007) "Thinking critically and creatively about focus groups," *Area*, 39(4): 528–535.

Hovorka, A. (2012) "Women/chickens vs. men/cattle: insights on gender–species intersectionality," *Geoforum*, 43(4): 875–884.

Hovorka, A. (2015) "The gender, place and culture Jan Monk distinguished annual lecture: feminism and animals: exploring interspecies relations through intersectionality, performativity, and standpoint," *Gender, Place & Culture*, 22(1): 1–19.

Joshi, S., McCutcheon, P. & Sweet, E. (2015) "Visceral geographies of whiteness and invisible microagressions," *Acme*, 14(1): 298–323.

Joy, M. (2010) *Why we love dogs, eat pigs, and wear cows: an introduction to carnism: the belief system that enables us to eat some animals and not others.* Berkeley: Conari Press.

Keith, M. & Brophy, J. (2004) "Participatory mapping of occupational hazards and disease among asbestos-exposed workers from a foundry and insulation complex in Canada," *International Journal of Occupational and Environmental Health*, 10: 144–153.

Latimer, J. & Miele, M. (2013) "Naturecultures? Science, affect and the nonhuman", *Theory, Culture and Society*, 30(7/8): 5–31.

Latour, B. (2004) "How to talk about the body? The normative dimension of science studies," *Body and Society*, 10(2): 205–229.

Law, J. (2002) *Aircraft stories: decentering the object in technoscience.* Durham: Duke University Press.

Lezaun, J. (2007) "A market of opinions: the political epistemology of focus groups," *Sociological Review*, 55(2): 130–151.

Longhurst, R. (1996) "Refocusing groups: pregnant women's geographical experiences of Hamilton, New Zealand," *Area*, 28, 143–149.

Longhurst, R., Johnston, L. & Hoe, E. (2009) "A visceral approach: cooking 'at home' with migrant women in Hamilton, New Zealand," *Transactions of the Institute of British Geographers*, 34(3): 333–345.

Lys, C., Gesink, D., Strike, C. & Larkin, J. (2018) "Body mapping as a youth sexual health intervention and data collection tool," *Qualitative Health Research*, 28(7): 1185–1198.

MacGregor, H.N. (2009) "Mapping the body: tracing the personal and the political dimensions of HIV/AIDS in Khayelitsha, South Africa," *Anthropology and Medicine*, 16(1): 85–95.

Mansfield, B. (2008) "Health as a nature-society question," *Environment and Planning A*, 40: 1015–1019.

Marres, N. & Lezaun, J. (2011) "Materials and devices of the public: an introduction," *Economy and Society*, 40(4): 489–509.

Miele, M. (2011) "The taste of happiness: free-range chicken", *Environment and Planning A*, 43(9): 2076–2090.

Miele, M. & Evans, A. (2010) "When foods become animals: ruminations on ethics and responsibility in care-full spaces of consumption," *Ethics Place and Environment*, 13(2): 171–190.

Neo, H. (2015) "Battling the head and heart: constructing knowledgeable narratives of vegetarianism in anti-meat advocacy" in Emel, J. & Neo, H. (eds.), *Political ecologies of meat*, New York: Routledge, pp. 236–249.

Pachirat, T. (2011) *Every twelve seconds: industrialized slaughter and the politics of sight*. New Haven: Yale University Press.

Philo, C. (1998) "Animals, geography and the city, notes on inclusions and exclusions" in Wolch, J. & Emel, J. (eds.), *Animal geographies: Place, politics and identity in the nature–culture borderlands*, London: Verso, pp. 56–67.

Pollan, M. (2006) *The omnivore's dilemma: a natural history of four meals*. New York: Penguin Press.

Porcher, J. (2017) *The ethics of animal labour: a collaborative utopia*. London: Palgrave Macmillan.

Potts, M. (2016) *Meat culture*. Boston: Brill.

Roe, E. (2006) "Things becoming food and the embodied, material practices of an organic food consumer," *Sociologia Ruralis*, 46(2): 104–121.

Roe, E. & Greenhough, B. (2014) "Experimental partnering: interpreting improvisatory habits in the research field," *International Journal of Social Research Methodology*, 17(1): 45–57.

Rose, G. (1995) "Geography and gender, cartographies and corporealities," *Progress in Human Geography*, 19(4): 544–548.

Saldana, J. (2015) *The coding manual for qualitative researchers* (3rd ed.). Thousand Oaks: Sage Publications.

Schreier, M. (2012) *Qualitative content analysis in practice*. London: Sage Publications.

Shukin, N. (2009) *Animal capital*. Minneapolis: University of Minnesota.

Solomon, J. (2002) *"Living with 'X': a body mapping journey in time of HIV and AIDS. Facilitator's guide,"* *Psychosocial Wellbeing Series*, viewed 14 February 2017, Available at: http://www.repssi.org/index.php?option=com_content&view=article&id=46&Itemid=37

Stanescu, V. (2010) "'Green' eggs and ham? The myth of sustainable meat and the danger of the local," *Journal for Critical Animal Studies*, 8(1/2): 8–32.

Sweet, E. & Ortiz Escalante, S. (2015) "Bringing bodies into planning. Visceral methods, fear and gender violence," *Urban Studies*, 52(10):1826–1845.

Temple, B. & Young, A. (2004) "Qualitative research and translation dilemmas," *Qualitative Research*, 4: 161–178.

Thompson, K. (2011) "Theorising rider-horse relations: an ethnographic illustration of the centaur metaphor in the Spanish bullfight" in Taylor, N. & Signal, T (eds.), *Theorising animals: re-thinking humanimal relations*, Leiden: Brill, pp. 221–254.

Vialles, N. (1994) *Animal to edible*. Cambridge: Cambridge University Press.

Whatmore, S. (2002) *Hybrid geographies: nature, cultures, spaces*, London: Sage.

Wilkie, R. (2005) "Sentient commodities and productive paradoxes: the ambiguous nature of human–livestock relations in Northeast Scotland," *Journal of Rural Studies*, 21(2): 213–230.

Wolch, J., Brownlow, A., & Unna, L. (2000) "Attitudes toward animals among African American women in Los Angeles," in Philo, C. & Wilbert, C. (eds.), *Animal Spaces, Beastly Places: New Geographies of Human–Animal Relations*, London: Routledge.

Wolch, J. & Emel, J. (1995) "Guest editorial on 'Bringing the animals back in'," *Environment and Planning D: Society and Space*, 13: 632–636.

Index

Pages in italics refer to figures and pages entries followed by 'n' refer to notes.

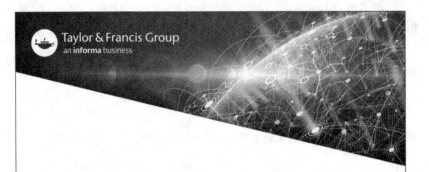

Printed in the United States
by Baker & Taylor Publisher Services